SHIPPING CONTAINER HOMES

*The Solution For An Affordable Dream-House. Discover How To Build A Shipping Container Home With This Beginner's Step-by-Step Guide. Best Plans, Design Ideas, Tips And **Extra Bonus Case Study** Included!*

ROBERT SAVINGS

© **Copyright 2021 – All rights reserved.**

This document is geared towards providing exact and reliable information in regard to the topic and issue covered.

- From a Declaration of Principles which was accepted and approved equally by a Committee of the American Bar Association and a Committee of Publishers and Associations.
In no way is it legal to reproduce, duplicate, or transmit any part of this document in either electronic means or in printed format. All rights reserved.
The information provided herein is stated to be truthful and consistent, in that any liability, in terms of inattention or otherwise, by any usage or abuse of any policies, processes, or directions contained within is the solitary and utter responsibility of the recipient reader. Under no circumstances will any legal responsibility or blame be held against the publisher for any reparation, damages, or monetary loss due to the information herein, either directly or indirectly.
Respective authors own all copyrights not held by the publisher.
The information herein is offered for informational purposes solely and is universal as so. The presentation of the information is without contract or any type of guarantee assurance. The trademarks that are used are without any consent, and the publication of the trademark is without permission or backing by the trademark owner. All trademarks and brands within this book are for clarifying purposes only and are owned by the owners themselves, not affiliated with this document.

TABLE OF CONTENTS

INTRODUCTION ... 6

CHAPTER 1.
WHY LIVE IN CONTAINER HOMES? ... 7

CHAPTER 2
WHAT ARE THE REAL PROS AND CONS? ... 9
Pros ... 9
Cons ... 11

CHAPTER 3.
THINGS TO AVOID AND TO PAY ATTENTION ... 13
Fixing Your Container on a Weak Foundation ... 13
Making Wrong Estimates as Regards Your Ship Size ... 13
Buying Containers That Are in Bad Condition ... 13
Paying No Attention to Codes and Building Regulations ... 13
Using the Wrong Kind of Insulation ... 14
Cutting Out Too Much Steel From the Container ... 14
Employing the Services of an Inexperienced or Unprofessional Home Builder ... 14

CHAPTER 4.
CONSTRUCTION OR PROJECT CHECKLIST TEMPLATE ... 15
1. Planning the Project ... 15
2. Finalize Design and Construction Approval ... 15
3. Site Preparation and Foundations ... 16
4. Place and Modify Containers ... 16
5. Finish to Occupancy ... 17
Final Thoughts ... 17

CHAPTER 5.
LAYOUT ... 18
Inspiration ... 18
Size and Layout ... 18
Positioning and Orientation ... 18
Floor Plan and Site Layout ... 18
Topography and Drainage ... 20
Access ... 20

CHAPTER 6.
DETERMINE A BUDGET ... 22

CHAPTER 7.
PERMITS AND ZONING LAWS ... 24
Container Home Building Regulations ... 24
Regulations and Building Zone Requirements of Some Countries ... 25
Tips to Get Your Container Home Permit ... 26

CHAPTER 8.
SITE PREPARATION ... 27
Deciding on Location ... 27
Site Work ... 28
Soil Types ... 29
How Do I Know My Soil Type? ... 30
Site Preparation Checklist ... 30

CHAPTER 9.
THE CHOICE OF THE CONTAINER ... 32
Importance of Container Condition ... 32
Right Time to Inspect a Container ... 32
What to Inspect and How? ... 33
Conditions and Container Grades ... 35
Why Choose New (and New-ish) Containers? ... 36
Why Choose Used Containers? ... 37
Cost of Shipping Containers ... 37
Shipping Container Condition to Buy? ... 38
Buying Your Containers ... 38

SHIPPING CONTAINER HOMES

Choosing a Container Dealer ... 39

CHAPTER 10.
FOUNDATION

Pier Foundation ... 42

Pile Foundation ... 43

Slab Foundation ... 43

Strip Foundation ... 44

Attaching Shipping Containers to the Foundation ... 45

Concrete Quality to Be Used for Foundation ... 45

CHAPTER 11.
CONVERT, RECEIVE, AND PLACE THE CONTAINER ... 46

CHAPTER 12.
HOW TO CONNECT THE CONTAINERS

Develop Layout Plans and Seek Spec Advice From a Licensed Contractor ... 47

Before the Joining Process, Modifications to Shipping Container Walls Should Be Made ... 47

Container Site Preparation ... 47

Putting the Shipping Containers Together ... 47

Tips for Connecting Shipping Containers ... 48

CHAPTER 13.
INSULATION

Cork Insulation ... 49

Spray Foam ... 49

Wool Insulation ... 49

Cotton Insulation ... 50

Fiberglass ... 50

Cellulose ... 50

Factors Affecting Choice of Insulation ... 50

Types of Insulation ... 52

Ways to Use the Different Materials for Insulation ... 53

Other Approaches to Control Heat ... 55

CHAPTER 14.
ADVICE ON ECO-FRIENDLY, ECO-SUSTAINABLE, AND ENERGY-SAVING SOLUTIONS ... 56

Appliances ... 56

Use Sustainable Resources ... 56

Recycle ... 57

Grow Food ... 57

Build Local ... 57

CHAPTER 15.
UTILITIES ... 58

Electricity ... 58

Gas ... 58

Sewer and Septic ... 58

Telecommunication ... 59

Water ... 59

CHAPTER 16.
ROOF AND CEILING ... 60

Roof ... 60

Installing a Ceiling ... 63

CHAPTER 17.
FLOORING ... 65

Inspecting the Original Floors of the Container ... 65

Remove or Not Remove the Original Floor? ... 66

Removing the Original Floor ... 66

Container Floor Replacement ... 66

Best Flooring for Flat Pack Shipping Container ... 66

CHAPTER 18.
INTERIOR ... 68

Doors and Windows ... 68

Installing Cabinets and Appliances ... 70

Adding Appliances and Fixtures ... 71

CHAPTER 19.

TABLE OF CONTENT

INTERIOR DESIGN IDEAS — 75
- Modern — 75
- Urban Chic — 76
- Rustic — 76
- Farmhouse — 77

CHAPTER 20.
EXTERIOR — 79
- The Style of Your Shipping Container Home — 79
- How to Install Cladding Onto Shipping Container Homes — 79
- How to Install Siding Onto Shipping Container Homes — 80

CHAPTER 21.
EXTERIOR DESIGN IDEAS — 81
- Cladding Materials — 81

CHAPTER 22.
HOW TO PAINT THE CONTAINER — 86
- The Kind of Paint to Be Used — 86

CHAPTER 23.
GENERAL ADVICE AND HOW TO SAVE MONEY — 87
- Careful Planning — 87

EXTRA
EXAMPLE CASE STUDIES — 89

FULL-SCALE FLOOR PLAN IDEAS — 94
- Plan 1 — 94
- Plan 2 — 95
- Plan 3 — 97
- Plan 4 — 98
- Plan 5 — 100
- Plan 6 — 102
- Plan 7 — 103
- Plan 8 — 104
- Plan 9 — 106
- Plan 10 — 108
- Plan 11 — 109
- Plan 12 — 110

FAQ — 111

CONCLUSION — 113

SHIPPING CONTAINER HOMES

INTRODUCTION

Building an actual house can be very strenuous and expensive, and, over the years, a solution was sought. There had to be a relatively cheaper house and easier to maintain than the usual ones available. For some other people, it was a chance to finally execute the brilliant ideas and designs that they had stacked up in their heads. So, something useful was discovered, and that is the cargo container used in transporting cargoes across the sea. The structure was just right for a house, and so, some very ingenious people thought about it being an actual house that people can live in.

So, gradually, with time, the shipping container home was implemented. It allows the chance of getting a house within a period of 3 months or less, depending on some factors that will be discussed later on in this book. This short time frame, as opposed to the years used in constructing the typical and traditional houses, was a big catch, and several people began to try it out. However, in as much as the concept of this sounds interesting, there usually are many things to consider.

Building a container home will require that you learn a couple of vital things so that you don't later have regrets. That is why this book was written. It contains pricey information you will need in constructing your shipping container home from scratch, the do's and don'ts, and all you'd need is a measure of readiness.

So, let's begin, shall we?

WHY LIVE IN CONTAINER HOMES?

CHAPTER 1.

WHY LIVE IN CONTAINER HOMES?

The concept of a shipping container home is relatively new to the world. They've become one of a whole host of options for people looking at alternative styles of housing. Many traces of the beginning—or a major uptick in alternative housing—can be seen after the 2008 financial crisis, where overpriced housing was very negatively exposed. Even in the next years, economic instability has loomed as wages have stagnated and income inequality seems to only grow. With these economic factors, it has spurred demand for alternative housing options. Shipping container homes are unique from the rest; therefore, they offer advantages that are hard to find elsewhere.

Shipping container homes offer several different advantages to potential home buyers and builders alike. In addition to potential cost savings, shipping container homes can positively affect the environment, offer rigid outer strength to protect from severe weather, and can often be moved even after construction of the inside is complete. Though shipping container homes definitely have drawbacks, their advantages are unique and not typically found in other types of housing.

Potential homebuyers who set out to construct a shipping container home by themselves may find that there are financial benefits to doing so. Builders can customize the home to only have the exact features they care most about. They can pick and choose where to find materials and utilize resources they know personally. A huge advantage to a low-budget buyer is that the container offers a very solid semi-complete structure almost immediately. The more fine details and luxurious upgrades can be made gradually as their budget allows.

Young adults today are more environmentally conscious than many of their older peers. With this in mind, shipping container homes are appealing for the way they can give new life to old containers at the end of their lifespan. Though this idea is a bit controversial, as containers can be scrapped and recycled, it may still be a beneficial option to reuse the containers in other ways. Additionally, some buyers will opt to make a best use of the container's strong metal shell instead of using more material to build exterior wall surfaces. The size of a shipping container also confines many homeowners to reduce the belongings they have, and thus, could theoretically lessen their environmental impact.

Depending on the desired location of a shipping container home, hazardous weather may be a huge factor. Shipping container homes are very rigid structures that can typically withstand extreme weather, including hurricanes, smaller tornadoes, and high winds. These containers are designed to withstand the massive waves and wind gusts while on board ocean vessels in rough waters. Though not always reliable if not properly maintained, they are designed to be watertight both to protect the cargo inside and in order to float should they fall overboard an ocean vessel. Shipping container homes can be a perfect concept for a low-cost housing option in a coastal area subject to hurricanes or typhoons or in the flatlands prone to tornado outbreaks.

Although mobile and modular homes offer similar advantages, shipping container homes can be moved even after construction is finished inside. Depending on the specific design, the homes can sometimes be moved to completely new locations, requiring only a new foundation and utility hookups. As stated earlier, the logistics costs of moving a shipping container home may be significantly lower due to not needing special oversize permits to travel down the road. Since shipping containers are so widely used worldwide, they also have much more equipment available for use to move and transport them.

WHAT ARE THE REAL PROS AND CONS?

CHAPTER 2

WHAT ARE THE REAL PROS AND CONS?

Living in a home constructed with recycled shipping containers has plenty of advantages, and more and more people are beginning to understand this and take action. You will appreciate these advantages first-hand and see them for yourself if you are lucky enough to have created your own and/or live in one.

When it comes to building your own home, used containers are versatile, sturdy, modular-friendly, and just an all-around winner, much cheaper than traditional construction techniques for timber-framed houses and brick-and-mortar structures.

It is now possible to use shipping containers for other purposes. These containers are now one of the best choices, considering the limited space and growing costs of building houses or offices. There were rumors that a branch of Starbucks was only made up of 2 containers. Looking at the interior design of the container, it looked like an elegant place to rest.

This makes them ideal for architectural homes as well. They deliver minimal costs for building material, plus can be easily changed. With 2 containers, you'll already have a nice location. While there were formerly prefabricated buildings, they were never intended as a place to conduct business.

Pros

Accessibility

The best price tag for building materials vs. steel containers will be the latter. You face financial difficulties each time you build a house; this will minimize the house's cost in so many ways.

Time-Saving

It takes time to construct a traditional building. From the moment the base is built to the end of the concrete supporting beams, there is no house or office. Prefabricated structures such as containers can easily become functional offices if you quickly need office space. This saves lag in the house-building process. The building of a metal container can quickly become an elegant room.

The average time to prepare and transport a shipping container is in the range of 2 months. This involves the time from the initial purchase to the complete personalization to create a comfortable home or office environment.

Furthermore, many companies are specializing in fitting containers for the fastest performance. It is nevertheless possible for those who want a more hands-on approach to fully complete the fit when the container is shipped and located.

Friendly for the Setting (Green Living)

So how green can vibrant shipping containers be?

Response: They can be as green as you would like them to be.

SHIPPING CONTAINER HOMES

Think of your home more like an "eco-pod" if you want to go down a very green path. By placing a few solar panels on the roof, you could generate your electricity. If you're close to a river or a quick-flowing stream, you can use hydro.

A "green/living roof" can be added to the top of your container to help separate and drastically reduce heating and cooling costs in the winter (in the summer).

Weather-Proof

Consider the fact that shipping containers on hundreds of thousands of miles of open-top trans-oceanic shipping container vessels are designed to endure the most unforgiving environmental conditions and constructed before being decommissioned to have a minimum work life of 20 years. After that, these containers have an almost endless lifespan in a fixed location.

They are made from prefabricated steel and welded, making them strong and rigid, and very durable. This makes them especially well-suited for high geologic activity areas, such as hurricane hotspots and earthquake zones.

Shipping containers can keep up to 175 mp/h (281 km/h) safely against wind speeds when anchored to pylons, which is easy to do. Whether it is constructed from used shipping containers or a conventional building, each building should have adequate foundations.

I listed the 3 main advantages above that most homeowners would note immediately. In truth, they are the 3 key things many people look at when buying any kind of home. It also reveals that the boom in shipping container homes has been entirely justifiable in the last few years and why so many individuals turn to these containers to create their own dream homes.

With the right internal floor plans, a comfortable and practical space can be created, with everything you need to relax in style.

Although it can sound like there are many difficulties in turning a container into a home, the whole process is usually straightforward. Once the secondary container has been purchased and shipped to your territory, it is simply a process of acquiring the correct plans for the intended purpose. The professionals will benefit from a high-quality finish. Moreover, experts would certainly need to install windows, doors, power supplies, and the like.

Excellent Flexibility

If the container has been purchased, the internal configuration and most essential features can be customized. It typically helps to look at various floor plans and floor sizes to understand what is available and create the layout to meet your particular needs. The traditional 20-foot container is an excellent way to turn it into a house since it is simpler to navigate and is better suited for combining it with other units.

The 40-foot model is a choice for those looking for more interior space and provides access to almost 300 square feet of space. Also, the wider container provides more versatility in separating internal space into different spaces.

WHAT ARE THE REAL PROS AND CONS?

Cons

Issue With Longevity

Many individuals wonder whether a hurricane or some other natural disaster would be deterred. Since it has no firm ground floor, at this point, the container is theoretically problematic as an alternative living space. Do people wonder if it is safe to live in this fabrication?

It is undeniable that rust and many other damages caused by elements deteriorate more rapidly as containers are made of metal. Why not spend your money on something durable?

Tropical Countries Not Advantageous

Heat is absorbed because the material is metal. Containers aren't very useful if you live in a tropical climate. These locations are not a practical place for construction because air-conditioning systems would increase your electrical costs. To adapt this form of construction to your home, you can now weigh the benefits and drawbacks of shipping containers used as prefabricated houses and offices. Although their popularity is now increasing, you need to realize they will not work in all circumstances.

They Need Proper Insulation

Regular insulation, commonly used in most homes, is rarely adequate for a shipping container home. You may require special insulation to enjoy a pleasant indoor environment. The insulation should be chosen following the climate in your area. If it's colder outside, you'll need to insulate to keep your indoor space warm. If it gets hotter, you'll need to protect yourself from the elements. As a result, when choosing insulation for a shipping container home, you must consider various factors.

Environmentally Harmful

Surprisingly, many shipping container homes are not as environmentally friendly as they appear. Because a container's original purpose was to transport goods, it isn't a suitable structure for habitation in its current state.

According to Ship Technology, concentrations of harmful chemicals or dangerous gasses in shipping containers exceed occupational hygiene limit values in 20% of cases. It can also be difficult to determine what gasses are present in a container, so having one tested before purchase is an important part of ensuring a safe living environment.

Limited Space

In the most basic models, custom container living embraces the concept of minimal living, with most homes having 320 square feet or less of living space. While combining multiple containers can create a much larger space of 1,000 square feet or more, the conversion process increases costs and reduces the project's sustainability. Those who choose to live in a container should be prepared to live in a smaller space or a container.

Expensive to Customize and Transport

While purchasing a storage container is inexpensive, transporting and making it habitable can be costly. The cost of transporting the container to you could be several thousand dollars, and you'll also need to own the land where you want to put it.

SHIPPING CONTAINER HOMES

Don't forget to budget for permits and prepare the site for construction, including adding public utilities to the property.

Spray foam insulation, wall reinforcement, wall removal, openings, and glass door installation, in addition to basic home finishes like flooring, plumbing, and roofing, all add up. While container homes can be assembled and built for less than traditional framed homes in most cases, they can also cost significantly more depending on the building's features, size, and design. Depending on your market, professional labor to install and finish a container home will cost $75 to $150 per hour. As a result, the total cost will ultimately be determined by the total square footage and the complexity of the finished construction. You can expect to pay at least $15,000 for labor, but more likely closer to $30,000 per container.

Could Bring in Health Hazards

Various chemicals, particularly insecticides, can be used to maintain some shipping containers while they are in use. Furthermore, the cargo loaded into these containers may have contained toxic substances. Sandblasting the shipping container to remove health hazards would be ideal. Toxic paint and spills can be removed this way, and the container can be repainted with non-toxic paint.

It Is a Real Challenge to Find Experienced Contractors

While it may seem easy to build homes with shipping containers, numerous issues can arise on the way. You should look for contractors who have worked with shipping containers before. Unfortunately, few contractors can properly handle such building materials, so this can be a real issue. Working with inexperienced contractors could lead to issues with your future home. As a result, it is worthwhile to conduct some research and hire the appropriate individuals for your project.

Structural Weaknesses

It is possible to modify metal containers to create the desired type of home, but it can be difficult. When large pieces of metal are cut out of a container's structure, structural issues can arise. To put it another way, the container will no longer be as strong and durable. As a result, you should follow a specialist's advice and use additional reinforcement if necessary.

Challenges on the Construction Site

Shipping containers will necessitate the use of a crane on the construction site due to their size. You won't be able to move them into the desired position otherwise. At the same time, there must be enough room for the truck that will deliver the containers. It should be able to maneuver and unload the container on your property with enough room. Also, before purchasing a container, it should be thoroughly inspected. Look for any damage or potential problems, such as dampness or rust. If a personal inspection isn't possible, request detailed photographs of the container. The images should be applied to the container's corners, doors, and hinges.

CHAPTER 3.

THINGS TO AVOID AND TO PAY ATTENTION

Container homes are one of the best housing projects to execute. However, in constructing your container home, the possibility exists that you could get distracted and make a couple of mistakes that could cost you greatly. Here we will study a few of the mistakes and then see how you can avoid making them.

Fixing Your Container on a Weak Foundation

Foundations, like the word implies, are the first things that should be constructed before any structure can be erected. So, what happens when you build on a weak foundation? Something about the ground is that it sinks with time. There could be a landslide or some other issue, and it gets worse when you have about 2 to 3 container homes merged together. That way, the container homes could end up getting pulled apart. You also should ensure that you construct your foundations out of strong components so that they can withstand the pressure of whatever you place in them. A good foundation works to spread the weight of the containers around the ground equally.

Making Wrong Estimates as Regards Your Ship Size

The containers used for shipping are not usually the same as those you would use to construct your home. Both of them come in different sizes, so the high cube containers are there to add extra inches to the height of your house. Insulating your container home will cut out of this space, and usually, what you end up having is a reduction in the general space available.

Buying Containers That Are in Bad Condition

You can buy different kinds of containers for your container home, but then, now is the time to consider the purpose they would be serving. Since you would be living in it, it has to be something of a quality that can stand the test of time. It has to be able to resist several factors like temperature, pressure, and weather conditions. Buying something that is already weak might need you to spend money on repairs later on. For example, containers meant for homes usually would have to be cut and welded at some point to construct doors and windows. All of these things work to reduce the structural integrity of the container. So, ensure that you properly scrutinize it before you get a container to prevent issues later on.

Paying No Attention to Codes and Building Regulations

Getting the right information is essential to avoid a scenario where your container home is dislodged from its original position. So, you have to ensure that you go through all the necessary permit laws of wherever you intend to put your container. Some countries, for example, are not too receptive to the whole container home idea, so you have to be careful. Not paying attention to codes and set-down laws could also lead to unnecessary time delays. So, ensure that you get the important information from whoever coordinates the zoning office of your area.

Using the Wrong Kind of Insulation

The type of climate that plays out in your area comes as a factor that hugely determines the kind of insulation that you use for your home. If you make the mistake of fixing the wrong insulation, you could suffer the risk of excessive heat or excessively cold temperatures inside your container. Some conditions like rain could cause the rusting of your container with time, and in a place where you have rainfall continuously over time, you could utilize insulation types like spray foam insulation. So, ensure that you make the right inquiries before working with any insulation.

Cutting Out Too Much Steel From the Container

A container can get really weak when you continuously cut out metal scraps from it. This way, your container could end up getting very weak, and consequently, it could lose its balance. So, what you should always look out for is a solid container home that can stand the effects of forces like wind. Even if you have to cut out steel scraps from your container, ensure that the cuts are made at a reasonable distance apart from each other. You could also try implementing steel beams into the structure of the container to support it.

Employing the Services of an Inexperienced or Unprofessional Home Builder

This is one of the biggest mistakes anyone can possibly make. So to ensure that your container home doesn't turn out bad, get someone who understands the technicalities of container construction and is very good at merging 1 or 2 things to get the desired effect. You have to ensure that the builder knows every important thing about every component, the uses, the types, the safe choices to make, and so on. The kind of work the person ends up rendering to you will largely depend on how much they know. So, to avoid a situation where you spend more than you originally planned for, ensure that you make the right inquiries before employing just anyone.

CONSTRUCTION OR PROJECT CHECKLIST TEMPLATE

CHAPTER 4.

CONSTRUCTION OR PROJECT CHECKLIST TEMPLATE

A construction timeline or schedule is used to visualize all of the tasks that must be completed during the construction process.

While numerous generic construction timelines are available on the Internet, none have been explicitly created for shipping container homes to our knowledge.

We're excited to share a shipping container home construction timeline because many of our readers have requested it.

1. Planning the Project

(4 to 6 weeks)
- Budgeting and financing (4 days).
- Create a preliminary home plan/design (20 days).
- Gather information and ordinances specific to the site (3 days).
- Materials to be used (2 days).

As you can see in the construction timeline, the planning stage will take approximately 6 weeks.
As you may know, planning is the most important stage of the project, and getting it wrong here can cost you a lot of money later on.

If you look closely at the planning phase, you'll notice that we've left out a stage for finding and purchasing land. This is because we assume you already own property. You must include this in your timeline if you do not already have it. It's difficult for us to give you advice on this because it could take days or months, depending on your personal preferences and the laws governing property transfers in your area.

2. Finalize Design and Construction Approval

(9 to 12 weeks)
- Complete your home's design and plans (10 days)
- Create construction blueprints (5 days)
- Decide on a final budget (2 days)
- Approval of local structural engineers (10 to 15 days)
- Permits for construction (15 to 25 days)
- Find a general contractor (3 days)

We've set aside 9 to 12 weeks for the second stage of the shipping container home construction schedule.

This is the longest build stage because it focuses on finalizing your design and getting your building permit approved.

As you can see, obtaining local structural engineering approval and obtaining building permits are the 2 most time-consuming activities.

SHIPPING CONTAINER HOMES

You're in luck if you're building outside of the city's zoning code or in an unregulated zone. You won't need a building permit, and the only thing you'll be doing at this point is finalizing your design.

Many people, however, will require a building permit.

It's best, in our opinion, to work with people and local governments who have experience with shipping container homes. If not shipping container homes, make sure they have experience with unusual construction—it will make the process go more smoothly.

The final task in this stage is to locate a local contractor to construct your home. Where you can find a local contractor is by far one of the most frequently asked questions.

Using your local networks is the best place to start. Do any of your acquaintances or acquaintances' acquaintances know anyone? A personal recommendation is always more credible.

3. Site Preparation and Foundations

(3 to 9 weeks)
- Identifying and purchasing shipping containers (5 to 15 days)
- Access, drainage, and land clearing (5 to 15 days)
- Layout and excavation of the foundation (5 to 15 days)

You will primarily be preparing your site for the arrival of your shipping containers during this stage.

To begin, clear the land and ensure that you have access to the location. Keep in mind that any contractors who arrive on-site will be bringing larger vehicles, particularly the trucks (and possibly cranes) that will be used to offload the actual container.

4. Place and Modify Containers

(3 to 9 weeks)
- Spray foam insulation and external air seal (1 day)
- Attaching containers to the foundation and putting them in place (2 days)
- Modifications, cuts, and reinforcements to shipping containers (1 to 15 days)
- Framing and roofing of structural elements (1 to 5 days)
- Put up new windows, doors, and siding (1 to 10 days)
- Rough-in for utilities (3 days)
- Framing inside the house (5 days)

Placing and modifying your shipping containers is the fourth stage of the construction process.

You have a few options for physically transporting the containers to your location, but they all require professional assistance.

CONSTRUCTION OR PROJECT CHECKLIST TEMPLATE

Once on-site, insulate the exteriors of containers before placing them in their final positions, which may be challenging to do. For the most part, we're just talking about the container's bottom. On the other hand, some people opt to insulate the outside of the walls and use corrugated metal for the interior walls.

Underneath the containers, spray foam insulation reduces heat transmission and prevents moisture intrusion.

5. Finish to Occupancy

(2 to 13 weeks)
- Insulation (1 to 10 days)
- Drywall (2 days) (optional)
- Fit flooring (2 days) (optional)
- Complete the plumbing, electrical, and HVAC work (2 to 10 days)
- Put the finishing touches on the fixtures, fittings, appliances, and trim (2 to 15 days)
- Decorating and painting (1 to 5 days)
- External cladding (between 1 and 10 days) (optional)
- Landscaping on the outside (1 to 5 days)
- Clean-up and final walk-through (1 day)
- Obtain final approval (1 to 5 days)

Your shipping container home should be dried in and finished at this point in the construction process. This stage may only take a few days if you're building a simple cabin-style container home. If you want to build a grand shipping container home with thousands of square feet of living space, meticulous landscaping, and a swimming pool, you'll have to wait months.

Final Thoughts

We hope that this shipping container home construction timeline has given you a better idea of how the process works.

It really is up to you and what you want to achieve.

SHIPPING CONTAINER HOMES

CHAPTER 5.

LAYOUT

Inspiration

Small homes are all around us. From far and wide, containers are being repurposed into personal residences. Whether your ideas are in the form of a houseboat to be used as a vacation home or you've always wanted to build that treehouse you've been dreaming of, this article can help get you started on your container home design.

In particular, will suggest how one design company went through with transforming an old shipping container into a home for their family. You'll see that the design work involved making modifications to the original container to make it more livable and comfortable for their family.

This is an example of "doing it yourself" and having a lot of fun building your own dream house. You'll learn that you can use containers not only to build your dream home but can also use them as a home office or a storage shed.

Size and Layout

- Trucks and trailers must not be longer than 24 feet.
- Your house is limited to a maximum of 8 feet wide but maybe up to 16 feet long.
- Your shipping container will have a door opening at least 8 feet from the end of its side, with your walkout porch on that side.
- You cannot exceed 200 square feet in total living space (including an open floor plan). This is based on exterior dimensions for standard-size precut containers.

Positioning and Orientation

Many people aren't aware of how important location and orientation are in terms of home design. However, being aware of this issue is critical to creating a practical and visually pleasant living area. To have a considerable quantity of natural light pouring into your house throughout the day, it's also critical to consider the views from different areas.

Floor Plan and Site Layout

Containers come in various shapes and sizes, which makes it difficult to accurately estimate what size and shape your container will be, but we'll work out a rough idea of the floor plans later on. This layout is going to be set out using a rectangle with around 8 feet ceilings (2 meters) to work out the different measurements for each wall:

- Width of container: * 2 + 18 inches (1 ½ meters) = Width of side wall - 1 foot (0.3 meters)

LAYOUT

- Height of container: * 2 + 8 inches (2 meters) = Height of ceiling wall - 1 foot (0.3 meters)
- Width of room: * 2 = Width of Each Room – 8 inches (0.2 meters)
- Room length: (Length of container - 1 foot) / 2 = Length of each room - 8 inches (0.2 meters)
- Room width: ((Width of container - 18 inches) / 2), then divide the answer by 2. Example: 9 feet 9 inches divided by 2 = 4 feet 4 inches, so that's the width for each room.

After that, we'll need to figure out how long the hallway and toilet are.

- Length of hallway: (Width of container - Width of each room) + 2 inches (0.5 meters) = Length of Hallway - 6 inches
- Length of toilet: This is taken by adding the widths, and then subtracting the width of a single door (1 inch).
- Width of toilet: * 2 + 18 inches = Width of toilet – 1 inch (0.3 meters)

Next, we'll take a look at different floor plans for your container home. Keep in mind that this is just a rough outline and should be used only as rough guidelines.

Area Calculations

Different layouts will require different amounts of area. For example, if you plan on building a 2-bedroom home, the amount of area you need for each bedroom could be as much as 80 square feet. This is based on the above measurements and considerations. Depending on your container's dimensions and desired floor plan, the amount of area needed can change drastically. We'll go through the various floor plans next, and explain how they use up different amounts of space per room.

- A typical home (2 bedrooms, 1 bathroom): Uses 34 square feet (0.36 square meters) for each bedroom and 22 square feet (0.24 square meters) for the bathroom and hallway.
- A large home (3 bedrooms, 1 bathroom): Uses 48 square feet (0.49 square meters) for each bedroom, 22 square feet (0.24 square meters) for the bathroom, and 21 square feet (0.22 square meters) for the hallway.

The Final Touches

Once you've figured out the plan of your container house, you'll need to add some finishing touches like a ceiling, flooring, and wall panels to complete it.

The simplest method is to use plywood or OSB panels, which are both inexpensive and simple to deal with. Plywood panels are preferable as they're more durable.

You'll need to cut the panels to fit each wall and ceiling. The easiest method to accomplish this is using a table saw; but if you're cautious, you could also use a circular saw or even a handsaw. The length of each panel will depend on its width and the desired height of the ceiling. The height of each panel won't be fixed, but rather depend on how tall you want your ceilings to be. For example, if your container's total height is 15 feet (4.5 meters) high, then your ceiling panel height should be 12 feet (3.6 meters).

SHIPPING CONTAINER HOMES

After you've cut all the panels, they'll need to be connected to each other. This is easily done by using some wood screws and some brackets made out of scrap wood. You should use at least 4.

Now that you've finished constructing the walls, floors, and ceilings; it's time to start thinking about insulation. Of course, adding insulation will depend on your climate and the amount of heat loss you're expecting; but if it's a cold area, then high-quality insulation is vital as it will save you a lot on energy bills in the long run.

Topography and Drainage

The container must be able to sit on a level and well-drained piece of land. This means you need either a small hill or good drainage in the surrounding area. It can also mean relocating existing landscaping that may be too close to the foundation of your new house. Make sure there are no trees close enough to stick their roots in your foundation under any circumstances.

If you want to prevent foundation difficulties, a basement is not a smart choice. Basements mean the container is sitting directly on soil and moisture.

For this reason, container homes are built with 4 walls and a wooden or concrete floor. This takes the load off of the bottom of the container so it does not soil as easily by having to support itself all the time.

It may be a good idea to have a crawlspace under your home as well. This would allow for plenty of room for storage for lawnmowers, shovels, and equipment that you may use in your yard without exposing them all to humidity from below ground.

Often, the future homeowner will not know how to properly maintain this type of home. This means your container may get moldy and wet, especially in humid regions.

So, cleaning out your container regularly is a good plan to keep rot at a minimum. When you are done with the container home, you need a place to store all of your junk from the old house. Be sure to move into or rent an indoor storage space when you are done with the container.

And finally, make sure you keep clean records for all of these materials moving into and out of your new house. A mail-in service can help ease any concerns about removing building material if needed in the future when it comes to selling your property.

Access

The shipping container home is a solution to the housing shortage in some of the world's poorest countries. Containers are durable and can be reused in many different ways. They can be converted into schools, clinics, or even homes. In this book, we will see design ideas for converting these containers into homes.

Some people want to travel more than they want to stay in one place, while others crave stability no matter where they go. There are many different reasons why someone might consider living out of a shipping container—accessibility being one of them. Many people have voiced their desire to live in containers, and several cities

LAYOUT

have instituted strategies to encourage this kind of living. Singapore is home to the "cargo village" concept, which features container-occupied homes that are designed for sustainability.

Perhaps you are interested in remote living and need a place to store your possessions and keep your things secure when you go out into the world. There are some options available for these needs: sea-turtle relocation centers at sea islands, underground storage units, or space stations. The latter option might not appeal to you, but it may appeal to some others.

SHIPPING CONTAINER HOMES

CHAPTER 6.

DETERMINE A BUDGET

Once you have determined that you want to construct your home from shipping containers, you will want to consider financing, budget, and insurance. These are big considerations that must be taken into account before the purchase of land and design. There are many things that will need to be included in your budget, and you will want to include them all before you ever purchase a piece of land.

Many factors will determine your shipping container home budget, such as:

- **New or used?** Will you be purchasing new containers, slightly used, or very used? The age of the containers can affect the cost.
- **What type of container and how many?** What type of container and the size are also big factors in the cost. As is how many containers you will actually need to complete your build.
- **Shipping.** You will need to determine how your container will reach the building site. Will it be transported by train and truck, or just truck? Will you need a crane to put it into place or stack it at some point in the build? Shipping can cost over $1,000 per container so be sure to factor it into the budget.
- **Can you weld and are you certified?** There are many points of the build that a welder may be required, especially when it comes to reinforcing the structural integrity of your home or adding reinforcement and structural support for a second floor. Alternatively, are you going to simply hire a contractor to do your work for you?
- **Who is going to be doing the labor?** Do you plan on completing the work yourself and what will you do about work that you are not qualified to complete? Will you be hiring contract labor at any stage or for the entire build?
- **Who will be pouring your foundation?** Home foundations have to be inspected and must follow a rigidly defined set of building codes; you may be skilled enough to pour one yourself, but it is highly recommended that you source this project out to save having to potentially re-pour a foundation and to ensure that it will pass inspection.
- **The budget for tools.** If you will be completing much of the work on your shipping container house yourself, then you may want to include all the tools that you will need to complete it. The list of tools that you will need to turn a giant metal box into a comfortable residence is quite extensive and may include many tools that are single purpose. The total cost of the tools that you will need if you do not already own them can add thousands of dollars to your budget. Tools that you will need may include a table saw, chop saw, numerous drills, an air compressor, framing gun, several different varieties of saws, hammers and mallets, tools for measuring and leveling, chisels, pliers, screw sets, bits, welding equipment, ladders, sawhorses, tarps, painting supplies,

DETERMINE A BUDGET

and safety equipment. You will need just about every tool known for housing construction, and that will definitely add to the total cost of the build.

- **If you will be living outside of the town water supply area, be sure that your property will be able to support a septic system and in some cases, a well.** This may require an inspection from an expert and proof of the inspection. Also, you will want to add the septic and sewer system to your budget. If you will be living far away from a city or county source of water, you will also need a well for the property for drinking water, bathing, and washing clothes. Check your county's requirements for septic system placement and well placement before you make any final decisions on the property.

- **Is there electricity already running on the property?** Depending on the location, confirm that you will be able to have electricity run to the property and the home once it is completed. There are some electric companies that may have restrictions about certain types of buildings or you may have to pay to install electricity on the property if it is in a remote location.

- **Will you be living off the grid?** If you will be living off the grid, then be sure to factor in key components such as a composting toilet or solar panels. You will also want to consider how you will get fresh water and how you will heat and cool your shipping container home. The steel that the container is made from is highly conducive to holding heat and will prove to be extraordinarily hot during the summer months without some type of cooling system in place. If you are living off the grid, consider your options for keeping cool and factor that into the budget.

- **Will you be including any additional features or large appliances**? Be sure to include patios, porches, washing machines dryers, garages, driveways, landscaping, or any other feature or item that will add to the cost into your build budget.

After you have included all the features and possible expenses for your build from the purchase of the land to the installing of the driveway, you will want to make a budget. Although shipping container homes are less expensive than traditional home construction, you may still end up with a home that may cost anywhere from $20,000 to $300,000 depending on the size, location, and features of your home. Your budget will depend on whether you are planning on selling the home afterward (and how much the profit will be) and whether you really need all those extra amenities.

SHIPPING CONTAINER HOMES

CHAPTER 7.

PERMITS AND ZONING LAWS

Shipping containers, whether you use them as a living space or as a shipping unit, are subject to varying laws and city codes.

You should also be aware that this is not just a matter of having approval from the local authority.

A ***container home building code or permit*** ensures that you have a building inspector accredited by the city to verify the integrity of your building plans. This inspector will examine the container's location, position, and positioning.

These checks are supposed to tackle a range of topics. Any of these safety questions are as follows:

- Surface rigidity. To guarantee that the soil will support the container's weight.
- The accumulation of snow
- Security of the structure
- Overturning or uplifting risk
- Adequate anchorage
- Resistance to wind

Depending on your place, the checks mentioned above could be subject to a different ranking format.

Here are some of the variables that could have an effect on permits in your state:

- Local ordinances and laws
- The container's actual location on the premises
- The planned construction's consumption time period
- The container's physical state
- The container's effect on the community/neighbors' attitudes.

So, what are the applicable container home permit conditions you should be aware of?

Container Home Building Regulations

Shipping containers, like conventional houses, must be designed in compliance with strict legislation and permits. To ensure proper licensing, make sure the structure is appropriately zoned.

Once you've been zoned, keep the following in mind, based on the city's zoning by-law.

Property Zoning

Zoning means the segmentation of the property into parts to specify the form of buildings that can be built. A government-controlled process; land zoning helps to plan the development of a metropolitan region. It is an approach that enables the government to group density-controlled related systems.

Zoning of property is important for any building project. It lets the state escape circumstances such as the repair of industrial buildings in particular parts of the capital, spirit shops in churches, etc.

Therefore, ensuring the right zoning laws in your area will influence the location of your shipping container.

Building Codes and Permits

The International Building Code (IBC) and the International Residential Code are also based on building codes and licenses (IRC). These uniform regulatory bodies have strict rules on power, fire safety, and plumbing inside buildings.

These codes are often revised every few years. Container homes can also be subject to different standards and licenses depending on the municipal authorities. So make sure you contact the local authority with the latest up-to-date policy.

Mobile, Modular, and Manufactured Building Codes

- **Manufactured homes:** Manufactured housing construction and safety standards are also known as mobile homes.
- **Modular homes:** Often built in factories and shipped for assembly to their planned location, these constructions are governed by the International Construction Code (IBC), not by the HUD.

Regulations and Building Zone Requirements of Some Countries

Each country has its own deal regarding regulations and building zone requirements. You'll have to factor in how many people are already housed within the lot and how high you are allowed to build. It'll pay you off in spades to be familiar with the zoning requirements for your region. Let's look into some of the countries where you are most likely to build a home and what you'll have to do to build in each of them:

UK

If you want to build in the UK, the first step is to get permission from the council. Check with the local planning authorities to find out what they have to say about planning and design. It's no good to build your structure just to have the council later tell you that you have to make major changes.

Any construction in the UK will require permission from the local council. The local planning authorities will each have their own specific regulations, so it is necessary to contact them before design and planning. From there, the council will be able to let you know if they need any more information.

Australia

Before getting going with a build in Australia, first, check out state requirements. Plan that into your checklist, and then move forward with the local council. Make sure that your build complies with both state and council regulations. You'll be able to get paperwork from them that legitimizes all building actions. Hit them up first, and you'll be set throughout planning and development.

US

Okay. The bad news is that if you're building in the US, you're going to need a building permit, for most places. You can get this by going to the local public works department. They can give you a full rundown of the zone your building falls in and the requirements needed for it. Just take this into account in your design, and you'll be good to go. You can set up your home to fit your needs and the building codes at the same time.

If you're especially lucky, you'll fall outside of zoning codes. Put simply, that means you can build however you want without a building permit. If you want, you can even select a site specifically because it's outside of the zoning requirements. Remember, though, that out-of-the-way sites will have restricted access to water, telephone, and power facilities. You'll have to factor this into the equation when choosing your building site.

Tips to Get Your Container Home Permit

It is recognized that it is never easy to obtain permits for your building. We have therefore built a few tips to make the process stress-free.

Here are some tips to get your *container home permit* faster:

- **The shorter the duration, the shorter the requirements.** The shorter you choose to keep a shipping container on your premises, the less demand you will have to fulfill. You do not even require a short-term permit in certain areas.
- **High traffic areas attract stricter rules.** You can face more permit rules if you choose to use your shipping containers in high-traffic areas. This is because some places are not associated with buildings. On the other hand, rural areas are not so rigid.
- **Hire a permit expeditor.** Your construction and the expeditor location will do you the same thing, based on the time you have to spare. This expert will help you save time by getting local authorities involved. They also do the paperwork and make enforcement quick.

SITE PREPARATION

CHAPTER 8.

SITE PREPARATION

Before your container is delivered to you, you need to make several preparations. Failure to plan this stage well enough could often result in spending a lot more than you originally planned. Most of the steps you take here aim to ensure that the site is well prepared to receive the container. So, to help you, we will discuss some things you need to cater for.

Deciding on Location

First, you need to decide where you want your container to be positioned. Which side in your land graph do you need it in? What factors should determine your decision? Let's answer these questions and more below.

Sun and Shade

The sun is one major factor that you must consider. On that cold morning, having the rays of the sun bathing your skin can be a very nice feeling, but then, at times when everything and everywhere is hot, the sun isn't what anyone wants. So, while observing your geographical area, ensure that you choose spots that feed in light softly to the insides of your container. You might also want to get an air conditioning system if you plan on setting your container in a sunny region.

Placing your container under a shady area or under a couple of trees will surely go a long way to help, but then, most times, these trees shed off their leaves during the fall and winter. To make calculations regarding the height of the sun at any place you are in, you could use tools like Sun Calc. It works at any date and time and helps you to consider implementing things like roof shades or lids for your windows in your project.

Topography and Drainage

This factor has to do with the shape and slope of your land. The slope is a factor that helps you to understand how water will run off it when there's rain. You surely would not want to find water puddles or sticky mud surrounding your container, as that could be quite messy. This factor will also help you see how you can change the direction of the water flowing across the land. Water, unfortunately, can be a home for mosquitoes, and snakes, so you have to exercise a lot of caution here.

Views

This section is where you think of what you can see through the windows and the doors of your container. Sometimes, the view of a hill, valley or moving water body will only add to the beauty of your container. So, while placing your container, ensure that you incline it in a way that allows you to have a good view of your surroundings.

Access

Think of how you will want to get into the container from the major roads. How long do you want the pathway to be? At what angle of elevation should it be? Should you include steeps or slopes? Do you need to get rid of any natural impediment from the way? Would you need to get rid of trees or other plants? Does the view get good or bad as you go up the road? You also would need to create a pathway that the building contractors and trailers can work with. It has to be wide enough and equipped with turns and angles that will make them have the best work output.

Site Work

This involves all the processes you will need to employ to ensure that your building site is ready for your container's placement. Here, you might need to follow a particular order so that you don't ruin one thing or the other. Let's discuss some activities that take place here.

Marking and Staking

This usually comes as the first thing that you need to do.

- Mark the ends that your container will be erected in.
- Mark the spots where the other things like utilities, roads, and buildings will be constructed on.
- For marking, you could just make use of paint or wooden sticks connected with ropes.

Clearing and Grubbing

- Clear the vegetation and debris off the land if there's any.
- Get rid of trees, tree stumps, rocks, if there are any.
- The more vegetation that covers the land, the more clearing you may need to do.

Grading, Cut, and Fill

- Here, you can begin by choosing first a foundation for your container. You could either grade the floor uniformly or go with the uneven floor.
- You could grade the land with concrete or raised slabs.
- Add swales and berms to shield your container home from the effects of a flood.
- Fix culverts, bridges, and low-water intersections at points where there is water flow.

Road Building

- Top the roads leading to your container with fillers like gravel, asphalt, or concrete.
- Avoid heavy equipment to prevent the roads from getting damaged.

Erosion Control

- A land that has been cleared of vegetation and dirt is usually prone to erosion.
- Erosion sometimes is caused by rain or the pile-up of sediments in ponds and streams.

SITE PREPARATION

- You can control erosion by planting useful vegetation near the area your container is to sit on.
- You could work with erosion control equipment like wattles and silt fences.
- Check the Environmental Quality Office of your state to get the necessary information as regards erosion and water pollution.

Fencing and Security

- Secure your container with container locks and grids.
- You could build a fence around the container.
- To build a fence, ensure that you mark out a line for it while you get things ready; building it before the container arrives would only limit the space for the dealers to work with.

Soil Types

When preparing your site, the first step is to check the soil composition and load capacity. You want to know that your site can handle the load of the home without tilting or caving in. So, if you know something about construction techniques, consider whether you want to go for pilings, piers, a trench, or a raft foundation. If you're setting up a crawlspace or room for a basement, now is when you need to consider it. The soil composition will determine what you have to do to set these spaces up to code.

All dirt is pretty much the same, yeah? Well, not really. Your soil might be sandy or filled with clay. It might be full of rocks or solid rock. The thing is, each of these soil types holds weight differently. So, you'll want to know what your soil type can handle before laying out all the plans.

Here are the soil types you're likely to encounter in your building site and some of the construction details associated with them:

Gravel

Gravel is a coarse-grained material that offers excellent drainage. It is relatively easy to dig out gravel to the desired depth and level it for a foundation. If you are working with gravel, then the best type of foundation is trench footing.

Rock

Rock can be a bit challenging to work with, but it's actually a blessing if your building site is a slab of rock. Just strip any surface soil and make sure to level the pad, and you're good to go. Rock has a high load-bearing capacity and can easily support a foundation. If the geotech recommends that you still use supports, the best option is to go with concrete piers. You'll first have to drill through the rock and down to the recommended depth. Then just frame it up and pour.

Sandy Soil

Sandy soil is composed primarily of fine-grained particles, often with some gravel and rock mixed in. It has a high load-bearing capacity if the weight is spread out over a larger area. Because of this, the raft foundation is the best way to go. Another thing to keep in mind is that it's best not to over dig the foundation, as this can place the weight on softer soil.

Clay

Clay is one of the less fortunate soil types you can encounter. It's extremely fine-grained and tends to hold water. The geotech reports will tell you how deep the clay extends. You'll have to dig down past the clay and backfill with more suitable soil. The best choices for the foundation are deep trench footings or concrete pile foundations.

How Do I Know My Soil Type?

The answer at this stage is pretty easy, though not the most fun for the DIY-minded individual. A geotech engineer could let you know what you have going on under your foundation easily. If you know that the site has some shaky soil, then the geotech will tell you how deep you have to dig for your foundation or pilings.

The way the geotech works is to perform soil bores every 100 to 150 feet. These are tests that describe the soil profile down to at least 20 feet through load-bearing soil. This will give you a solid understanding of the load-bearing capacity of the space. It will also tell you about the drainage capacity of the soil. You can get detailed reports about density, water content, particle size, and soil classification. After the geotech assessment is done, you'll walk away with a comprehensive understanding of the soil beneath your site and what it can handle.

At the same time, you'll be given assessments regarding elevation and surface qualities. This will let you know how best to level the site, if it's necessary at all, and how to work around challenges with the site and soil. Here are a few things you'll find in a typical geotech report:

- Surface soil type
- Subsurface soil type
- Soil bearing capacity
- Optimal soil compaction
- Recommended foundation type
- Recommended foundation depth
- Drainage requirements
- Groundwater depth
- Frost depth

Tip: Sometimes the local authority has information on the soil profile of your area, especially if you plan to build within city limits. If you plan to build something a bit further out, you can almost guarantee that a soil profile will be necessary and should be factored into your budget.

Site Preparation Checklist

- Get a geoengineering soil profile.
- Determine the most appropriate foundation type and depth.
- Meet with a structural engineer to design the foundation.
- Decide if you can pour the foundation yourself or if you need a contractor. If you do need a contractor for the job, hire well.

SITE PREPARATION

- Calculate the concrete strength required for the job.
- Calculate the volume of concrete required to pour the foundation.
- Dig the foundation.
- Place the molds.
- Place a gridwork of rebar and steel wire into the molds to reinforce and offer flexural stability to the concrete once it has cured.
- Mix concrete to the necessary strength and pour it into the molds, ensuring that the foundation is level.
- Pour with your weather conditions in mind, ensuring that the concrete can maintain a stable temperature as it cures.
- Allow the concrete to cure for 5 to 7 days before placement.

SHIPPING CONTAINER HOMES

CHAPTER 9.

THE CHOICE OF THE CONTAINER

When you start to consider building your container home the most important step you can't avoid toying with is getting the right shipping containers. They serve as the walls, floor, roof, and general structural support. Do not approach this decision stage with the mind that containers are all the same as long as they serve the purpose. You need to understand that containers vary in quality, price, and conditions. You need to be assured that you are paying for the right quality and working conditions you need before structuring your budget.

The size and the quality of the container should be one of the first things you should consider. Will it allow you to use different types of aesthetics? What is the performance level? Is it even with the cost? Is the cost affordable and equivalent to your estimated budget?

Condition is very necessary for consideration when selecting a container. When you select a container with good working conditions, it will be easier to plan your shipping container home and select a workable design. However, a poor container will cost you more and reduce the effect of all your aesthetic designs. Also, you have to draw your design, as it is a guide to helping you select containers that have the specifications you need.

There could be a temptation to buy low-quality containers and to reduce your cost. However, you've got to be conscious of the fact that fewer quality containers could require more maintenance investment and expenses to put them in place so as to withstand harsh weather conditions.

Importance of Container Condition

You have to determine what quality of containers you need to help you adequately set your chosen building project. Purchasing the right container makes building much easier, as you have fewer requirements and the need for modifications. Your design will help you determine the size of the container that you need, and the quality of the container you need to get.

Right Time to Inspect a Container

As you commence your container home construction, endeavor to take the time to inspect the container yourself. Failure to exercise this duty of care will require that you have to readdress resultant problems and address them at certain points. These can cost incredibly much when repair issues come up. Inspections could be tiring and complex. However, you have to take the pain to inspect these metals before making investments. You can make investments without necessarily making detailed measurements and conditions for more specifications. There are 2 basic types of inspections. Before we go into discussing them, let us consider the specified details. Your inspection choice is dependent on where you purchase your container. You might not get the chance to inspect your container before delivery. This doesn't mean that inspections are different and the major difference is the time of inspection and the specification of inspection required.

THE CHOICE OF THE CONTAINER

Pre-Purchase Inspection

Inspecting before purchasing your container is best especially if you are close to the location or vendor you will be purchasing from. It is wise you take the pain to go and see it instead of making an order or sending a random individual to get it delivered to you. If the distance is too much for you, you can make a video or photo inspection of the container. Give specifications on how you want the video to be done or the photo to be taken. A live video will be most appropriate.

Post-Purchase Inspection

After the order has been made and the container delivered, you need to check the container to be sure it is in the right shape with the specifications you ordered for. You have to be present to validate your container after it arrives or get someone trusted to do that for you. You don't just receive it and keep it or start working with it without validating its state of arrival. Depending on the mode of operation of the company delivering the containers to you, you'll have to sign off on the delivery. You might not be allowed to inspect the interior on arrival but if there is something wrong with the container, you'll discover it from the exterior. The sooner you discover an issue and report it, the better.

What to Inspect and How?

When setting off to launch your container home, you must take adequate time to do a thorough review of your shipping container yourself. Overlooking this step can expose you to certain dangers in the long run. Although, some contractors could work with lower quality containers and build supplementary designs.

You can adjust your design to allow you to modify damaged areas and use them to make aesthetic decorations or make necessary cuts for windows or doors in that area. The kind of inspection you are required to carry out doesn't involve any special skills, except consciousness and precision in observation. You can create a procedure for carrying out your inspection so you get to miss nothing.

When inspecting visually, scan the container with your heads and eyes in the same direction preferably a direction perpendicular to the path you are working. If you are inspecting a corner scan walking back to front and keep your eyes moving up, down. You can use recording tools like a long selfie stick that will allow you to take videos of locations and parts, your eyes cannot assess. Another tool you'll need is a flashlight and a ladder to give you full visual experience, and a hammer to check for rust.

Structural Frame

The strength of a container is based on the 12 steel beams situated at the 6 edges. These beams come in different sizes dependent on the rail structure. Damage on the beam is almost unfixable. They tend to contain deep cross-sections. Hence, to get a view of their true state you'd have to assess it from inside and outside. Your concern should be on deeper rust because it is a sign of corrosion, which reduces the beam's strength. Surface rust poses little or zero risks.

Underside

This part of the container is invisible to the eye up front but it is made up of small beams that stretch through the container length. This is where you screw your plywood into. Because this part is permanently hidden from the sun which should dry

SHIPPING CONTAINER HOMES

it when it gathers moisture, it is almost impossible to prevent it from rusting. However, you could take advantage of the time in between when the container is being moved from the delivery vehicle to the mapped-out layout portion, and take snapshots. If you miss this opportunity, you might have to hire equipment to lift the container up. Nonetheless, damage in this area can easily be repaired without hampering your aesthetics.

Walls

A container has 4 rectangular walls that are very much visible to the eye for inspection. The container is enveloped in corrugated steel that helps to sustain its structural stamina. Surface rust can be a challenge, if it is too severe it will be hard to cover up. Also, if there is a see-through hole on any side of the walls, it could be a sign that the container is really weak, and wider holes are bound to come out at a time. However, you can take advantage of the damaged work areas to make door and window cuts. Also, note that dents that are robust enough to be visible can interrupt your container designs. Container walls are usually deep, when a dent goes deep enough to be visible on the side, there is a situation.

Roof

You can inspect the top of the container with your selfie stick or use a ladder to inspect. You should be on the lookout for fair conditions like signs of stagnant water, old patches, and dents. You can also use the hammer test here but be careful to not put your body weight on an area that has been weakened by rust. You can only tell if the roof is waterproof when you check the interior.

End Doors

This is the only access point to your container. Note that part of a structure that moves might get weak and become obsolete when affected by dirt and corrosion. Check the hinges, all 4 lock bars, and the lock, observe how they spin. Also, check the perimeter of the doors for missing parts that need to be replaced.

Interior

This is a continuation of the exterior inspection to validate all that you saw on the outside and ensure that the interior is waterproof. Check the walls and roofs with your flashlight to enhance your view. To find out if there are openings in the walls, close the odors and feed your eyes on how to fix them. Use spray pipes to confirm the waterproof feature. Hence, you can patch your roof with the right materials once you identify faults.

Floor

This is another section you need to pay full attention to. The most permeable floor material commonly used is plywood. This means that it easily absorbs toxic chemicals.

Documentation

When securing your shipping container, certain documentation is important or necessary to acquire. One of them is discussed below:

- **CSC plate:** In some locations, it is also regarded as Combined Data Plate and it is the most detailed part of the container. This data is on a single metal plate

THE CHOICE OF THE CONTAINER

that is permanently glued to the container. Under this CSC plate, there is specific information that is worthy of note.

- **Container identification number:** Also termed BIC Code, this number is built together with an equipment identifier, check digit, serial number, and owner prefix. It is an identification number that points out the owner and the vehicle. The check digit helps you to verify the correctness of the container identification number.
- **Type code:** This code is made up of 4 distinct digits. In some cases, it is a mix-up of letters and codes. For instance, 42G1 means average height, general-purpose, and 40-foot container.
- **Classification society approval:** Certain societies approved the container before it got to you, this number includes a reference to this approval organization.
- **ACEP/PES:** The CSC plate will definitely have enough space for an ACEP number and follow-up PES examinations. This ACEP number is usually in correspondence with the classification society approval code.

Conditions and Container Grades

The highest standard that validates containers is the CSC certification. Besides this, there is no other validation. Hence, it could be difficult to grade a container adequately. There are different certification and validation grades. Some intra-category grades have been created for containers by certain companies, however, they cannot be trusted because these categorizations were placed on one side by one person's perception.

When assessing a container that is placed on sale and has been categorized by an expert, you are expected to get an understanding of the seller's position in determining the condition and grade of the container. While you examine different sellers' classifications and grading, you need to understand why it was made so in the first place. You can sign off a deal that will allow you to regain your money if the container doesn't meet your specifications. Also, note that these categorizations are based on aesthetics and cosmetics in place of structural implications. When assessing lower quality containers, you need to determine if the wears and slight damages affect the condition of the container's structure. Not all sellers use grades, so we'll divide the container types into 2. They will be discussed further below:

New (and New-ish) Containers

These are brand new containers that are either one-trip, refurbished, and clean new. Refurbished is a tricky term to use here as no standard covers for it. In some situations, it is also categorized as used. You can categorize it yourself after understanding to a balanced level the steps involved in the refurbishment.

One-Trip Containers

This container is sometimes likened to the category of new containers. However, this container is commonly used to move a single cargo load from the manufacturing company in a country to another country. After fulfilling this service, these containers are packed up for sale. You can bet that it is as good as new. There are only complications when there are situations of mishandling. In many locations, you are certain to get one-trip containers for small money.

Refurbished Containers

Refurbishment is a word that describes bringing the old to new or dead to life. It must have gone through a process of repair, restoration, and review. However, refurbishment is also limited to aesthetics restructuring. Over the years, container owners have stopped refurbishing due to the excessive cost required. Many of them just decide to replace, as it is a much cheaper option. Refurbishments are now conducted by local sellers or resellers, not users. Sometimes, refurbishments could make a container not be in correspondence with the CSC certification because it might still have some poorly patched areas and replacement parts. The difference between a new and refurbished container is quite easy to tell. Their performance which is the main center of focus cannot be decided except upon verification. Refurbishments might include the removal of rusts, dents, scratches, repainting, and priming. It might also require the replacement of doors and revamping an entire wall. However, the fact remains that it is much affordable to get a good container than trying to refurbish an old one and make it look new.

Used Containers

These are containers that have exhausted a large part of their useful life. There are usually categorized by barely used and completely used up. Containers do not basically have expiry dates but if used for a very long time, they can start to wear and tear, and eventually become risky for use at the time of depreciation. It makes no sense to invest at this point as the wear and time would never end until the container is condemned. Most containers in this category are those that have been repaired or refurbished at a point. However, the repair can only hold it together for a long while before it all rusts and become useless.

Cargo Worthy (CW) Containers

These containers are categorized and certified by CSC certificates. This categorization implies that the container has a performance standard of a high percentage when compared to the details in the original specification. They might not be appealing to the eyes but they are as strong or almost as standard as new containers. They are confirmed to be in better shape than most used containers. However, modifying their exterior by cutting would make the certificate invalid.

As-Is Containers

These are less standard containers when compared to cargo-worthy containers. The As-Is have many cosmetics and aesthetics defects that it is fairly encouraging to purchase. When purchasing this container, you would need to be on-site to see the nature and level of defects and determine if it is good enough for your project. This container is known for different types of wear and tear, and it is never sold on warranty, so the risk of danger or accident upon purchase is fully borne by you. However, it could still prove to be a really good bargain.

Why Choose New (and New-ish) Containers?

- ◆**Uniformity:** When you purchase a couple of these containers for your building project, it is easy to get them set up together without having to bend or

THE CHOICE OF THE CONTAINER

make many cuts for alignment and uniformity's sake. You would save a lot of money you would have spent on modification, painting, removal, and repairs that would most certainly come with purchasing a used container.

- **Appearance:** They are appealing to the eyes, and attractive to behold. You can keep them in your yard without raising awkward stares from nosy neighbors.

- **Life span:** You have a sure guarantee for long useful life before modification is required.
- **Peace of mind:** Although used containers can be less disturbing and more intensive after modification, the assurance that your container is in good shape gives you rest and peace of mind that you can use it to set up any project that you intend. You can set up any house structure of your choice without fear of unforeseen dangers.

Why Choose Used Containers?

- **Affordability:** This is the major reason 90% of the persons who bought used containers do so. Used containers pose as a cheaper alternative to acquiring a shipping container for whatever purpose.
- **Availability:** You can find a used container very close to your geographical location without much stress because most companies happen to use and dump shipping containers when they are done. Getting a new or high-quality container might require you to travel for long hours to the port to get one for yourself. You can find your container early and start setting up your building structure early.
- **Eco-friendliness:** Reuse is always beneficial to the environment and helps to keep the environment safe because you are not gathering materials to rebuild. You are building with already existent materials; this is called adaptive reuse or recycling. In some developed states, you might get a certification for green building or tax evasion incentive to appreciate your goodwill.

Cost of Shipping Containers

Prices of shipping containers differ based on condition (new, new-ish, or used), type (height, length), age, and location. Nonetheless, the prices are always closely related. We have computed an average price level or bracket for getting a shipping container in any location or quality of your choice. It is not a sure price, remember it is only an estimate. You might get your container for a cheaper or more affordable price.

- Used standard 20 feet container: $2,100
- Newish standard 20 feet container: $3,000
- Used standard 40 feet container: $2,850
- Newish standard 40 feet container: $5,600

SHIPPING CONTAINER HOMES

- Used high cube 20 feet container: $2,200
- Newish high cube 20 feet container: $3,200
- Used high cube 40 feet container: $2,950
- Newish high cube 40 feet container: $5,800

Shipping Container Condition to Buy?

Now that you have come to understand the different container conditions and their implications, as well as their pros and cons, you should be able to decide on the container that is best for you and the structure of your project. Here are a few questions to help you select your container condition and make a purchase decision:

- **Will you need to move your container?** If your answer is yes, you should be on the lookout for CW containers.
- **Do you intend to build in large windows and doors?** If yes, carefully engage in inspection before purchase to avoid any weakness on the main support beam.
- **Will you be opening a whole part of the container walls to allow for larger open space**? If yes, ensure upon your inspection that the main support beams are in a good shape.
- **Will you use a separate corrugated sheet for roofing?** If yes, you should be less bothered about damages discovered on the roof of the container.
- **Are you particular about cosmetics and aesthetics?** If yes, then you should consider a new container or budget for refurbishing a used container to your taste.
- **Do you intend on replacing the floor or modifying the container floor to more standard?** If yes, the damages on the plywood floors should bother you less.

Buying Your Containers

When you are deciding on the quality of shipping containers, it could be difficult making your selection when you are ready. Prices could be scary and you don't know how to trust the result of your inspection. Also, different sellers deal in shipping containers ranging from dealers, resellers, distributors, and middlemen, and all of these persons have their own processes. However, once you have decided finally on the condition and quality of the container you want to get, the work is made easier. The next decision stage is deciding where you want to buy your container.

Most business purchases are now done online. The Internet is now a full-fledged marketplace. Hence, you can focus your search online. Finding a container offline is not impossible but it could be difficult.

Online—In the US

If you live in the United States, check out BoxHub. This is a website that allows you to find containers nationwide, conduct your inspections, evaluate pricing levels, and make your complete purchase. In certain circumstances when you happen to be less satisfied with the order upon delivery, they provide a refund policy. They have

THE CHOICE OF THE CONTAINER

a team that works in direct relations with specific shipping companies across the nation and stand as middlemen to help them sell off the containers to individuals or retailing companies. They tend to deliver quality at a lower cost when compared with other companies. They also coordinate delivery constructively at a very affordable rate. It is proven to be an easier way to purchase your container online in the US than using Amazon and many sophisticated seller options. If you are unable to connect with BoxHub, you can check out other companies on the Internet that are also keen on quality and affordability. You could also use your city code to search on google. Make a search like "Shipping container + London." You can use the name of a particular city in your state for a more narrowed search. If you are settling for striking the purchase deal with an individual or firm that is not registered as a container dealer, you might have to travel over there for physical inspection to validate its quality and arrange for delivery logistics afterward.

Online—Outside the US

No matter your location, every country has access to shipping containers. There is surely a way to track down a container that you need, either on a seaport or container terminal. You can find seaports using the Sea rate website. You will find several seaports alongside information about their operations. You can inquire from there if they sell containers and their price ranges. If you leave too far from the coast, a land port option is best considered for you. You can check out options close to you using Craiglist or Gumtree. Most commonwealth companies have major ad sites for trading containers. Ensure to conduct a broad search before settling for any decision. Take the time to scroll through all available results and validate the results with the condition, quality, and cost requirements you have. One-trip containers can mostly be secured in Asia, you can get in touch with container dealers from that region through Alibaba. Although, it is more costly and stressful, as you have to do a lot of paperwork and inspections but you are sure of getting a high-quality container.

Finding Containers Offline

There are a lot of container dealers in space that you might not actually find until you start searching. Some businesses and individuals just have containers sitting around their vicinity that they don't know what to do with it. If you notice any container lying around, you can look for the owner and strike a bargain for purchase. Some of them might be pretty glad at the prospects of clearing their yard for money. This is an incredibly good catch you wouldn't want to miss, so keep your eyes open.

You can take advantage of personal referrals from people you know in the shipping or logistics industry. If you contact any dealer within your location who failed to help you strike a deal, ensure to ask for a connection to another until you find the container you want.

Choosing a Container Dealer

Choosing a container dealer is almost as important as choosing a container because when you set out to buy a container, you are establishing a business relationship. Also, there are a lot of container dealers out there, you want to be guaranteed that you are dealing with the right one.

SHIPPING CONTAINER HOMES

Integrity and Reputation

In selling transactions, the term knowledge asymmetry is used to describe the fact that most sellers are more knowledgeable than the buyers as regards the subject of the transaction. Hence, you have to find a trustworthy seller that won't look for ways to play on your ignorance and rip you off your money for a low-quality product. You can easily evaluate the trustworthiness of a seller by checking out their membership in organizations in the industry like Container Dealer's Association, Intermodal Association of North America, National Portable Storage Association, and many others. This doesn't totally validate a seller but it gives validation to a level of trust and faith.

You can also consider finding out other buyers' opinions through buyer's reviews online or inquiries from people in your area who have made certain purchases from a container dealer. Finally, you can also establish trust by verifying your purchase. At the point of price agreement, take note of the CSC plate information and after the container is delivered do a check to confirm if it is the one you selected.

Warranties and Returns

Demanding a warranty and return allowance upon dissatisfaction is a good bet you can bid for. Note that it is only a trustworthy seller that would consider giving you a warranty offer. Some warranty offers come with extended fees on rare occasions. If you don't want to make the payment, you can forgo the purchase altogether as it could be a red flag sign. Although warranties could be invalid when you are ordering a shipping container to build a home, ensure to read up the warranty offered before making the payment for the purchase and ensure to understand the terms and conditions applied to avoid unpleasant stories.

Volume Discounts

Discounts are only applicable when you are purchasing more than one container. It is wise to buy all the containers you need from a certain vendor to enable you to bid for a discount and makes your record-keeping easier. In some cases, the seller openly discloses them and in other cases, you might have to demand it with a negotiation before making payment. Discounts are only applicable to volume pricing which is an important concept in business marketing. Here are 2 things you need to be educated about before striking off a discount negotiation:

- **Average Customer Value (ACV):** What is the revenue or turnover rate a single customer generates for the company.
- **Customer Acquisition Cost (CAC):** What are the required expenses a company incurs in getting a single paying customer.

You might not have the numbers; a good estimation is enough. If you are buying many containers, you are saving the company's ACV in getting a new customer. Hence, it is only reasonable that the company reduces your purchasing expenses by giving you a discount. If you are good at negotiation, you might strike a good bargain and save yourself some money.

Value-Added Services

There are regular sellers and value-added sellers. Regular sellers just sell off their commodities to you. Value-added sellers help you to put the commodity to use after selling it off to you. They might charge you for rendering these services because

THE CHOICE OF THE CONTAINER

they are necessary but it is worth the extra penny. Some of the extra services that are common in container sales are delivery and offloading. Others are modification and set up which includes, painting, installation, insulation, and welding. You have the option of letting your container dealer do it for you or hiring a contractor. Before you settle for any of the decisions, make inquiries and find out the quality of service rendered by your container dealer, and compare it with your contractor's work with their prices.

Delivery and Offloading

When selecting a container dealer, look out for those that can help you coordinate the logistics or you might have to engage in finding out companies that can assist in logistics. Large container dealers possess necessary equipment for delivery and offloading that smaller companies or dealers might not have. Once the delivery is set, be cleared on the processes, the responsibility of offloading and security of container during transit. Another factor to consider is the time of delivery. Anyone that is trustworthy can deliver a product but the question is, "when"? How long will it take to get 10 containers delivered to your geographical location? Although most sellers cannot give you an accurate forecast because business is dynamic and sometimes the future is not really measurable and could come unexpectedly, they might not get orders early to deliver alongside your order. Just get familiar with their terms of operation.

SHIPPING CONTAINER HOMES

CHAPTER 10.

FOUNDATION

There are **4** primary types of foundations that are used for shipping container homes—pier, pile, slab, and strip. Well, there are some other types of foundations; however, these are the most commonly used ones for shipping container houses.

Pier Foundation

For several reasons, pier foundations are the most widely used alternative when it comes to shipping container homes. They are quite cheap, easy to build, and are DIY-friendly as well. Pier foundations are made of concrete blocks. In order to properly optimize the concrete pressure, each of the concrete piers or blocks uses comes with 50 centimeters x 50 centimeters x 50 centimeters and steel-reinforcing on the inner side. When it comes to shipping container homes, concrete piers are generally used at each corner. You can use 2 more piers that can be mounted halfway down on either of the container if you use the large 40 feet containers. You will get the chance to save a great deal of money and time on the walls of the piers. Also, you will not need to dig a huge amount of ground for installing the piers.

All you need to do is dig the soil for the piers, typically 50 centimeters x 50 centimeters x 50 centimeters in dimension. In comparison to a baseboard, the entire space under the shipping container needs to be excavated. Pier foundation is by far the most famous base for shipping container homes. It is also the most recommended one out of all.

FOUNDATION

Pile Foundation

When the foundation of the soil is too thin for supporting the concrete piers, pile foundations are put into use. It is the costliest one out of all. Piles, which are cylindrical rigid steel tubes, are pounded into the soil by the soft soil right before the piles meet a better load-bearing soil. Once the steel piles are properly secured in place, they are capped using a concrete block. They make the ground look the same as it would look if concrete piers are used. Such a type of foundation is not something that is DIY-friendly. As special tools are needed for installation, getting a contractor for installing the pile foundation is better.

Slab Foundation

A slab foundation is a common alternative when the ground is soft and permits even distribution of weight. However, building a slab foundation is quite costly and time-intensive when compared to building a pier. If you want to use a base plate, you will need to prepare yourself to dig a lot. Concrete slabs are used over which you can place your shipping containers. The base in this foundation is marginally smaller than the home footprint. For instance, if you install 2 containers of 40 feet, the base will be 18 feet wide to 42 feet. It will have a somewhat overhanging basis all around the perimeter of the shipping container. One of the primary advantages of this kind of foundation is that there will be no empty gap in your home base. So, it can be said that a slab foundation can provide your house with a stable backbone.

However, slab foundations are not inexpensive due to the additional mortar required as well as the large amount of space that must be excavated. It can also be used in cooler conditions where freezing is not an issue. But as the temperature of the surface falls below the interior temperature, the risk of heat loss also increases. It is because the shipping containers can easily conduct heat to the soil, which can move heat in comparison to convection into the air. Keep in mind that access to the service lines will be limited when concrete is attached to the slab foundation. Whenever there is a leakage of water lines, you will have to replace the concrete for providing access to the leaking lines. If you use a pier foundation, you can also access the power lines.

Strip Foundation

Also known as a trench base, it is a mixture of slab and pier foundation. The strip foundation is a surface of the concrete that is laid for holding the containers. The band of the strips is generally 4 feet tall and 1 to 2 feet wide. The strip can either run all around the circumference of the container or can also be placed at the top and bottom of the shipping containers. It is a suitable option when you want something less expensive for the foundation but has a smaller base. A rubble-strip foundation of loose stones right under the concrete strips can be used in all those areas where the ground soil is quite damp because of a huge amount of moisture. Also, these stones help water to flow and drain properly. But in comparison to any of the other foundation options, strip foundation is the most fragile one of the lot. As they come with a shallow shape, the strip bases are only suitable for small to medium constructions.

It is the most important factor for a shipping container home to have a stable, clean, and solid base. In strip foundations, there are 3 types of used foundations that come with different cost rates. The first foundation is a trench floor. It is constructed of brick and block masonry. It is filled with concrete. A base of somewhat $5,000 can easily accommodate larger containers. The second one is a pier foundation. A pier foundation of $5,000 is enough for holding a big shipping container on a site where the soil is firm. The last one is the slab foundation which is the costliest of all. It will

FOUNDATION

cost you around $6,000 as it uses both steel and concrete bars for stabilizing the home.

Attaching Shipping Containers to the Foundation

The most widely used method of connecting the shipping containers to the pad of foundation is by using a steel plate. The cast alternative includes pushing down a steel plate with anchors onto the wet concrete. Once the anchors are mounted, you will have to epoxy them in place. You can also use mechanical anchors; however, they are not that necessary all the time. Also, they are less powerful in nature and are not that much recommended. In any of the cases, you are looking out for a flat level of a concrete plate for complementing the 4 fittings of the corners on each container. If the concrete has properly dried or healed, you can set the shipping containers on steel plates. Now, you can solder them both together. There are people who tend to place the shipping containers directly on the floors. In such instances, tremendous weight is deposited. It is more or less alright; however, you are also needed to be aware of tornados and flooding that can easily drive a loose container.

Concrete Quality to Be Used for Foundation

When someone decides to go for a concrete box, the next big question that arises is the strength of the concrete that needs to be used. Geotechnical engineers determine the concrete strength that you need to use for your home. The strength of concrete is known as the C rating. The most widely used all-purpose concrete, concrete C15, consists of one part of cement, 5 parts of gravel, and 2 parts of sand. The greater the amount of cement used, the stronger the concrete. For instance, C30 is a super-strong concrete that includes one part of cement, 3 parts of gravel, and 2 parts of sand. If you need to mix tiny amounts, it can be done using your hands. You can also opt for a cement mixer.

For instance, if you need to find out the required amount of concrete for 10 feet deep and 22 feet long base plate, multiply 22 x 10 x 2. The quantity of concrete that you will need is 440 cubic feet.

If you put concrete in very hot weather, make sure that you prepare the site before pouring the concrete. You can place temporary shades to block all the direct sunlight. Also, you will have to spray the ground with cold water right before pouring the concrete. Another great tip is to pour concrete either early in the morning or late in the evening to prevent excessive temperatures.

SHIPPING CONTAINER HOMES

CHAPTER 11.

CONVERT, RECEIVE, AND PLACE THE CONTAINER

You can now prepare to receive your shipping container once you have set a budget, picked a location, finished working out a design and layout, obtained all requirements, and prepared the area you will be building on.

Purchasing a shipping container is the first step you must accomplish before you can move on to constructing your new home. Container homes that are smaller and simpler might cost anywhere from $10,000 to $35,000, while large residences with many shipping containers and amenities can cost anywhere from $100,000 and $175,000 to build. The price of each container will be determined mostly by its size and condition. Newer containers are more expensive, but they last longer and have fewer complications when it comes to construction. Containers that are used or older are less expensive.

When it comes to transporting your shipping container, the costs may vary depending on the size of the container, the distance it will travel, and the cost of insurance. Transporting a shipping container locally is much cheaper than transporting internationally—it can cost around $1.5 to $3 per mile if shipped locally. However, the costs can change considerably if the shipping container is from outside of the country. Most shipping containers are imported from Asia, which means that transportation costs can be quite costly—delivering a shipping container overseas might cost anywhere between $2,000 and $20,000. While the price will still be determined by the size, distance, and cost of insurance, purchasing a shipping container in your local area is significantly less expensive.

Once you are ready, you can then receive your shipping container and begin construction on your new house. So long as you have properly prepared the area where you are placing the shipping container, then you will have no further worries.

HOW TO CONNECT THE CONTAINERS

CHAPTER 12.

HOW TO CONNECT THE CONTAINERS

People from various industries and requirements use a variety of methods to connect shipping containers together. We're frequently asked about welding them together and stacking container limits. We'll go over how shipping containers are frequently connected in the sections below.

Develop Layout Plans and Seek Spec Advice From a Licensed Contractor

Have a detailed plan or design in place before you get too caught up in creating a clear-span space with your shipping containers—this includes future add-ons. This type of planning ensures that everyone involved in the process is aware of the project's requirements and specifications.

It also allows you to consult with or hire contractors who can help you maintain the integrity of your shipping containers as you begin to modify them for assembly.

Before the Joining Process, Modifications to Shipping Container Walls Should Be Made

You'll need to modify the walls first if you're working with several standard shipping containers. Beams and bracing columns can be installed in a tasteful manner as needed and with future additions in mind. You should also make sure to neatly frame any openings you've made in your shipping container walls to help create a uniform and strong seal between containers.

You might also want to think about the benefits of buying an open side container because their large doorways are already framed. To create an open space, simply remove the doors.

Container Site Preparation

Make sure both units' foundations are compacted and level. This will help keep your structure looking nice and uniform and create the best deal possible.

Assemble the first unit and place it on top of the foundation. Before starting to position the second unit, make sure the first unit is completely straight and in place.

Place the second unit in place and go over your measurements once more. It's critical that everything lines up perfectly because changing the position of the shipping containers after they've been connected is extremely difficult.

Putting the Shipping Containers Together

Plates and caulking are one of the most common methods for connecting shipping containers and ensuring that the seam between them is properly sealed.

Tools required:
- Drill
- Threshold plates
- Screws
- Flashing

- Caulk

You can start the connection process now that your shipping containers are connected by placing threshold plates along the seam. Make sure to use screws to hold them in place. Interior threshold plates can be installed along the floor.

On the outside of the container, threshold plates and flashing (roof) can cover the seams on the walls and ceiling.

Caulk, Roofing Cement, or Welding Can All Be Used to Make a Good Seal

Finally, most people will use caulk to help seal the container against the elements and improve efficiency. A ridge cap, along with additional roofing cement and roofing materials, can help protect the roof of your new 2-container building.

Is it Possible to Join 2 Shipping Containers With Welding?

Yes, indeed. This is a more permanent way of joining 2 shipping containers together if you have welding experience.

Can You Stack Containers Vertically?

Yes, but make sure that your weight is distributed primarily through the corner posts so that the foundation takes the brunt of the load. We strongly advise using structural engineers and architects when constructing a multi-story shipping container building to avoid catastrophic collapses.

Tips for Connecting Shipping Containers

When connecting 2 or more shipping containers to form a container home or business, you'll almost certainly need to add a few new support columns on the inside to ensure structural integrity. This means you'll need to make room for it in your interior design plans.

What makes you think you need more help? Once the 2 shipping containers' roofs are perfectly welded together, they will theoretically be joined. This distributes more weight than the initial design of either shipping container anticipated. When snow or rain piles on top of the roof, a few interior support columns will ensure that it does not buckle or bend.

Many contractors will build small wall segments around the columns in the center of your container if you're concerned about their appearance. This gives the building a more welcoming appearance while also providing more space for insulation materials.

INSULATION

CHAPTER 13.

INSULATION

Insulating your shipping container home isn't really a luxury. Whether you live in a hot or cold area, shipping containers can easily transmit heat (or cold) and make your life miserable! There is also the problem of drainage and the possibility that water might leak into the container, whether from its roof or the underside. This is why you have to take insulation seriously and work hard to ensure that your container home is well-insulated. That way, the weather won't be a problem and won't affect your daily life, neither will underwater sources or even rainwater.

There are different approaches to insulating shipping containers, and it is trickier than you might think. The main challenge faced when trying to insulate a shipping container is how thin the walls are. Yes, they are sturdy and durable, and they will carry the loads when it is time, but those container walls are also pretty thin, which makes insulation complicated. There is a way around this, but it might mean taking up some of the container's interior space.

Another factor you need to consider while looking for a suitable insulation material is how you plan on building your walls. You will go with different approaches depending on whether you will be adding several containers together; if so, the necessary room for insulation becomes less of a problem. This won't be the case if you're making a snug single or double-container home. In those cases, it is possible to add exterior insulation.

You can use different materials to insulate your shipping container home, whether on the outside or inside.

Cork Insulation

Cork is natural insulation that provides good results. The great thing about cork is that it is renewable and a natural source that is biodegradable since it comes from trees. More important, you don't have to cut down the trees to get the cork. Another significant feature of cork insulation is its acoustic properties, forming an acoustic buffer in your home that will stop sound from leaking outside or coming in from outside. This is particularly important for shipping container homes because those thin steel walls can easily leak sound.

Spray Foam

Spray foam is one of the most popular approaches to insulating containers, and it is one of the fastest ways to do it. The great thing about spray foam insulation is that it's applied to your container's interior and exterior walls. This is useful if your container has been coated with paint that can sometimes have toxic organic volatile compounds added to help the steel survive long periods in the sea. With spray foam insulation, you can contain such compounds and stop them from spreading into your home.

There are different types of spray foam insulation. It is generally a good idea to invest in the best available because it can prolong your home's life, protecting you from several things. Icynene is generally considered one of the best options for spray foam insulation. It is a water-sprayed foam insulation that uses tiny plastic bubbles to fill the insulation's interior, providing excellent insulation and protection. It also doesn't have as many organic volatile compounds as other spray foam products, and those that are there can disappear after only a few weeks.

Wool Insulation

This is one of the natural approaches to insulation, and it also yields good results. Wool insulation is renewable and completely natural, seeing as it comes directly from sheep's wool. This insulation is environment-friendly and quite efficient, providing powerful insulation comparable to denim, fiberglass, and other fibrous insulation options. Another great advantage of wool insulation is that it naturally contains lanolin, which is a flame retardant. This means you don't have to treat the insulation with other chemicals for fire protection.

Carefully consider your options before you purchase wool insulation because some types are better than others. Look for companies that sell wool insulation and research the different varieties they offer before settling on a particular type.

Cotton Insulation

Cotton is another natural source of insulation that is environment-friendly and efficient. An advantage that cotton offers is that it can be recycled from other clothing sources, so you don't need to source new cotton; pretty great for the environment! Like wool, cotton provides excellent insulation comparable to fibrous insulators like fiberglass. Like wool, boric acid (a natural fire retardant) is usually added to cotton in commercial denim, which means you don't have to treat it for fire protection. The downside to cotton is you have to make sure it doesn't get wet because moisture causes it to lose some of its insulation properties.

Fiberglass

Fiberglass is made from superheated sand, and, in other cases, recycled glass spun into thinner fibers. It is cheap wall insulation that is also pretty efficient, which is why it is very popular in many countries.

Cellulose

Cellulose is a loose-fill insulation that relies on adding macroscopic materials in the walls' cavity. The chunks of the insulating material are added, but for this insulation, the wall cavities need to be completely contained, or else the material will just spill on the floor. Cellulose is made of recycled paper products that get shredded and then blown into the cavity using a specialized machine.

Factors Affecting Choice of Insulation

Choosing insulation for your home is a major step in this construction process, and you need to take your time and do it right. Insulation is integral to keeping your home at a moderate temperature compared to the outdoors. Each type of insulation has its own pros and cons, and you need to consider the advantages and disadvantages before picking a certain type. These are some factors that might affect your choice:

- ◆ **R-Value:** This industry term refers to thermal resistance per unit area. It is basically a number that expresses how well a material can prevent the transmission of heat. For instance, cotton and wool have an R-value of about 3.5 per

INSULATION

inch, which is good. But spray foam has an average R-value of 3.7 per inch—even higher with certain varieties. As you can see, the values differ, and this is a number you need to consider while selecting insulation.

- **Performance:** The performance of the insulation isn't just affected by its R-value. Other factors come into play, like the open or closed-cell structure of the material (for open-cell foam R-value is 3.2 to 3.7 per inch, while for closed-cell foam it's 6.5 to 7 per inch), entrapped gas, and others. These aspects affect performance characteristics, and you need to consider each before investing in a particular type of insulation.
- **Air leakage:** Good insulation should be able to stop air from flowing through it or around its edges.
- **Cost:** As with the rest of this shipping container home project, the cost is something that you must consider. This doesn't just include the materials cost, but also labor and equipment expenses if you won't be able to do it yourself with your tools at home. For instance, spray foam insulation's average cost is around $0.5 per board foot for open-cell spray foam and $1 to $2 for closed-cell spray foam. If you're having professionals install it for you, their time will also be factored in the expenses. Still, spray foam is considered one of the more expensive options compared to the rest. Cellulose costs $1 to $1.3, fiberglass is $0.64 to $1.2, rockwool is $0.9 to $1.65, cotton is $0.76 to $1.4, and wool is $1.33 to $2, all per square foot.
- **Ease of installation:** How easy is it to install this insulation? If it is easy, then you can DIY and save money on labor and equipment. If it is too complicated, you will need help. While the obvious choice is to save money, your shipping containers might need a special type of insulation that will require outside help. Blanket insulation is generally considered the easiest to install, and it is available in fiberglass, wood, and fibers. On the other hand, spray foam is not as easy and is not recommended for a DIY approach because it requires experience and skills, so you'll most likely have to hire someone to do it for you.
- **Net interior space:** This refers to how much space remains in the interior of your container after applying the insulation—if you applied it on the inside.
- **Vapor permeability:** Can vapor flow through the insulation? How well does the insulation prevent the vapor from seeping inside and lingering there? Materials like fiberglass, wool, and cellulose are considered semi-permeable, while mineral wool is a retarder, as are most foam types, except cementitious foam, which is considered vapor-permeable.
- **Sustainability:** We mentioned earlier that some insulation types are more eco-friendly than others, which is an important factor to consider for many people. A lot of shipping container homeowners choose sustainability for minimal impact on the environment, so the insulation's sustainability might be a factor to consider.

Types of Insulation

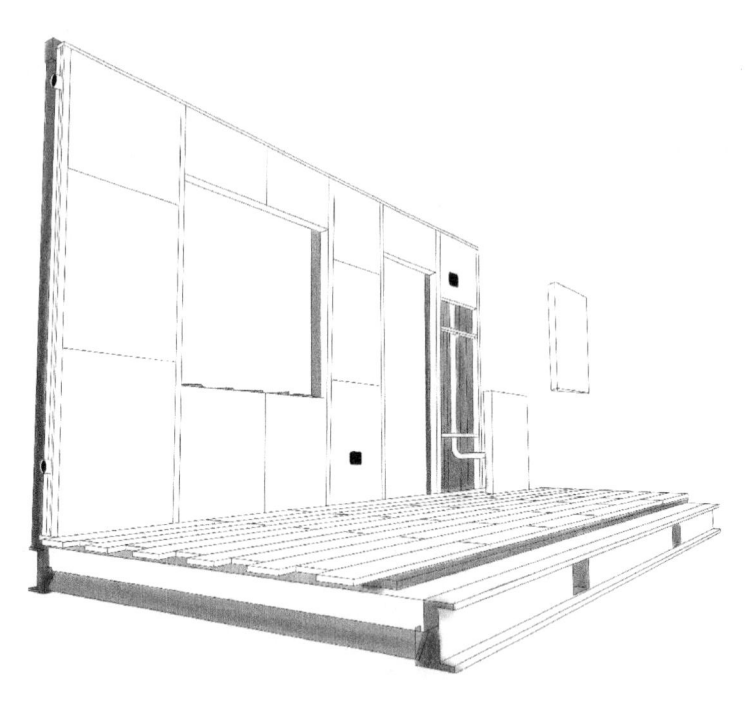

Wall and ceiling insulation application.

When you think about insulating your shipping container home, you have to think about what approach you want to follow—interior, exterior, or both. Considering that shipping container homes are basically metal boxes, they are excellent conductors of heat. So, the best approach will be to insulate both the interior and the exterior of the container for the best results. This is especially important if you live in extreme weather conditions where choosing just one type of insulation would lead to heat control problems in your home.

External Insulation

The concept of external insulation is simple. If you don't have it, the container will easily heat up. Relying just on the internal insulation will lead to heat or cold seeping in through the internal insulation, affecting your entire living situation. This applies to both summer and winter, and things will be much worse if you suffer from extreme seasonal changes where you live. In short, external insulation will help keep the home cool in summer and warm in winter. This also means it will reflect on your energy expenditures; you'll save on heating/cooling costs with proper inside and outside insulation.

Another cool feature of external insulation is that it can help improve the outer facade of your shipping container home. One approach that some people follow is filling the voids of the corrugated container wall with insulation, spray foam most likely, and after that, will be ready for paint or cladding. This is, however, a somewhat more expensive option that might not work with all budgets.

It is also important to insulate the container's underside because a lot of heat and moisture might seep in or out from there. The best time to do that is when placing your container on the foundation. If that doesn't work, you have another option of adding insulation underneath the flooring, which we will discuss later. In any case, make sure there is some form of insulation on the underside of the container.

INSULATION

While we recommend insulating both the interior and the container's exterior, some people prefer saving inside space by only doing exterior insulation. Not modifying the container's internal walls does preserve a lot of floor space, but, you need to make sure your external insulation is done properly. External insulation will help you preserve that floor space while providing some heat control. Remember that you will also need to insulate your container's roof, whether you are leaving the original roof or adding a new one.

If you are adding a new roof, adding spray foam or other insulating materials underneath it should be fairly easy. If you are leaving the roof as it is, you need to cover it in an insulation layer. This is particularly important if you don't plan on adding a ceiling inside the container since this means you won't be adding the interior insulation that comes with the ceiling—this one doesn't take up any extra height. In other words, if you don't plan on adding a ceiling, make sure the roof is well insulated.

Internal Insulation

Many people ignore internal insulation, thinking it is not really essential, but it can make a world of difference. While external insulation does the biggest job in controlling the heat or cold seeping into the shipping container, climate may still make it past that first layer; this is where internal insulation comes in. To partition and frame your container's interior, then adding insulation won't take up much space, and the same goes for when you add a new ceiling. Suppose you leave an exposed ceiling (the container's original). There, you should expect rust after a while since there will be a lot of condensation inside the container and affect the original steel ceiling.

The great thing about spray foam insulation inside your house is it can help you improve the place's aesthetic value. It can cover up any dents or scratches or any other marks on the walls, and it is easy to paint over spray foam insulation. You can use external insulation or internal in certain places or you can double up in areas such as the roof/ceiling to minimize any heat seeping into your home.

Ways to Use the Different Materials for Insulation

We discussed earlier the different materials you can use for insulation, and now, we will list the different ways you can use these materials. Each type of insulation has its pros and cons. You be the judge of which works best for your home.

Blanket/Roll Insulation

This is considered the cheapest available insulation. The most commonly used material with blanket insulation is mineral, which is also known as rockwool. Installing blanket insulation inside your shipping container home requires stud walls. You should know that the rockwool rolls are put between the battens and then you roll them down in place. As long as you have the stud walls in place, adding in blanket insulation should be simple and straightforward.

Know, though, that blanket insulation is made out of fiberglass, so you need to treat it with caution to avoid damaging it. Always use protective gear, including masks, goggles, and gloves. Compared to insulation types like spray foam, blanket or roll insulation is considered somewhat cleaner, but it is more time-consuming than spray foam and a bit more complex.

SHIPPING CONTAINER HOMES

As we just mentioned, to install blanket insulation, you need to have stud walls in place. After that, you will simply place the blankets or rolls in the gaps between the studs. The great thing about this approach is that you can insulate the wall without cutting anything, which reduces the time needed for this process and is generally less wasteful. Still, plan ahead and calculate the battens and blankets' width so you can do this with no cutting. Since you are using blanket insulation, it is always recommended to put the foil against the container's wall.

While it is recommended to use spray foam insulation at the underside and top of the container, you can use blanket or roll insulation—but you will need to add battens first. So, add the beams across the width, whether that is on the top or bottom of the container, and space them at a distance of 1.5 inches between centers. Some people mix different methods, using spray foam and blanket insulation together. You just put the panel or blanket on the foundation or roof and add a thin layer of spray foam.

Spray Foam

We talked earlier about spray foam and how it is one of the best insulation approaches—and the fastest. However, spray foam insulation is also a bit tricky and requires some skill and experience so you can install it properly. It is also somewhat messy and can cover places you don't want to be covered with foam. This is why it is recommended that you always cover such areas while working with spray foam. Be careful to cover pipes, wires, windows, doors, electric sockets, and any other area or item in the shipping container you don't want covering in spray foam. Using a cut-to-size plastic sheet is an effective way of protecting those areas. Also, remember also to cover the floors with plastic before you add the interior insulation so don't clean up any mess later on. Pro tip: Cover cables and sockets and pipes with tape before spray foaming.

One great thing about spray foam insulation is that it does not need battens, which means you can spray it directly on the walls and save a lot of time otherwise spent making the battens. It is also an excellent insulator that does a better job than most insulators out there. This is because spray foam provides an airtight barrier and the best heat resistance per thickness among other insulators. Another great advantage of spray foam over blankets is how easily you can use it to fill gaps and even out uneven surfaces, unlike panels or blankets which need to be cut to fit into such confined spaces.

You don't have to frame the inner sides of external walls to insulate with spray foam; if you do, it will allow you to install plasterboard or paneling over the insulation. Plasterboard will offer an even and smooth surface you can paint, and panels also have their own special look that can be left as is. If you follow this approach, spray foam into the space between the battens like you would be if you were using blanket insulation.

Using spray foam insulation, you should add at least a 2 inches layer of foam thickness to the wall. You can spray foam the full 2 inches on one side of the wall, or you can divide it between the interior and exterior walls by adding 1 inch on the inside and the second on the outside. Remember that spray foam can seal any gaps that result when you join 2 containers—around bolts and joints, and adjoining floors.

Panel Insulation

This is your third option. It is a bit more expensive than blanket insulation but cheaper than spray foam. It is also one of the rather less complicated means of insu-

INSULATION

lating walls in your shipping container. One advantage is that it is quickly set up, and it is not as fragile as fiberglass. Panels are somewhat thin, but they do provide very good insulation. You can get them in fixed sizes and place them between the studs as you did with blanket insulation, and the installation process is pretty much the same. Consider the fact that panels are thinner, so you will have more space to work with, unlike blanket insulation. You can use panels to insulate the container's underside, too, just like blanket insulation, but you need to add battens to the foundation to affix the panels.

Other Approaches to Control Heat

You will notice we have discussed insulation and controlling the shipping container's heat even before talking about handling the container when delivered. This is because insulation or maintaining a container's temperature are critical aspects you need to keep in the back of your mind even before receiving the containers. However, there are other approaches to tackle the temperature situation inside your containers that don't include insulation.

Plants

While you must plant trees, flowers, and plants around your home, that doesn't mean you can't do the same inside or on your shipping container home. One of the best ways to control the heat inside shipping containers is to invest in greenery. You can create a green roof—a garden on the roof of your home with plants and grass. While the greenery won't work as an insulator, it can reduce the thermal radiation coming in from the sun into your house. You can also do this even if you have insulation because a green roof is an excellent idea whether it is aesthetically speaking or to supplement your insulation with fresh air and reduced heat.

You can implement other techniques in the design of your house, like passive heating and cooling design. In this approach, you design your container home so it dampens the energy needed to heat or cool the place. This can be done using techniques like solar chimneys and Trombe walls, among others. This approach is great in moderate climates, but if you live in an area where there is constantly high temperature and scorching sun, it won't be sufficient on its own.

At the end of the day, insulation is your best answer for controlling heat in your shipping container home. There are a ton of options, and even the limited types of insulation we mentioned have subcategories and derivative compounds. While insulation is critical to your home's design, you shouldn't dwell on it for too long because there are other pressing matters to attend to. The important thing is to make sure that insulation works within the grand scheme of your design, choosing a type that will work for your particular home.

SHIPPING CONTAINER HOMES

CHAPTER 14.

ADVICE ON ECO-FRIENDLY, ECO-SUSTAINABLE, AND ENERGY-SAVING SOLUTIONS

Interior design aside, you need to think about ways to improve your shipping container home's sustainability. One of the most important reasons many people invest in such homes in the first place is how they can be used to reduce carbon emissions of a residential home, not to mention live off the grid with minimal damage to the environment.

Contrary to popular belief, while container homes are better than traditional ones in terms of environmental friendliness, not all shipping container homes are necessarily eco-friendly. Your practices determine just how sustainable your shipping container home can be, and there are things that you will need to do to make sure that your shipping container home is environmentally friendly.

Appliances

Electrical appliances around the house are one of the areas in which you can seriously reduce your carbon footprint and minimize emissions. For starters, we leave a lot of devices on or on standby, which results in wasted electricity and all the harm that comes with that. It is obviously easier for us to leave devices on standby mode to turn on quickly, but it is wasteful. This is why a better approach would be turning these devices off when you are not using them, whether that is done by unplugging the machine or turning off switches.

You might think that standby devices don't necessarily consume too much power, but you would be mistaken. Studies show that at least 10% of residential electricity consumption is done through standby devices, which is huge. So, reducing energy consumption and being sustainable is definitely worth waiting a few extra seconds or even minutes until the device powers on.

Moreover, if sustainability is something you care about, you should also start considering investing in eco-friendly appliances, and there are many options. You can get an Energy Star-labeled refrigerator that minimizes electricity consumption in your shipping container home. You could also consider switching to a gas oven instead of an electricity-operated one, which would save you a lot of money on utility bills.

It would also be great if you replaced all your old appliances with new ones that run on more efficient systems that can reduce water and electricity consumption. Eventually, you will find that you not only created a much more sustainable living environment, but you also saved a lot of money because of those practices.

Use Sustainable Resources

The good news is there are always a lot of sustainable resources out there that you can use to make your shipping container home truly sustainable. The catch is, you might need to make a little more effort to find those resources, but the results are definitely worth it. You can start by using eco-friendly insulation. We talked earlier in the book about eco-friendly insulation like cork and wool or cotton, and you have other options like straw and hemp. Depending on your design and the climate in which you are building the shipping container home, it will be much better for the environment if you can use such materials for insulation.

Another resource you might change is your energy source. A solar panel system is ideal for a shipping container home living off the grid. You might pay a few thousand dollars to install the photovoltaic cells, but once that is completed, you will save a lot of money on energy bills in the long run. Solar energy is also one of the most eco-friendly solutions out there to generate electricity.

Recycle

Living in a shipping container home is sustainable in its own way, and if you want to really take things to the next level, consider recycling. This doesn't mean you should give away items to be recycled, you can do it at home. You should obviously have a garden around your shipping container home, and there is no better way to maintain that garden than using compost made out of everyday items we usually dispose of.

So, make a compost bin, in which you can make a compost pile to be used in your garden. Things like fruit leftovers, vegetable peels, cotton clothes, tea bags, paper, and a lot more can be used in your compost as a way of recycling these items. When you recycle this much garbage you would usually throw away, it is much better for the environment. Such items would usually get tossed into a landfill, increasing the buildup of methane gas.

Grow Food

Picking up from the last point, and since you will be recycling your garbage, why not grow your own food? You already have a garden, so you absolutely can grow vegetables, and this practice can save a lot of carbon emissions. Plus, making your own food means you control what is added to the soil, so you know there are no added chemicals, and you get to enjoy organic and fresh vegetables.

Build Local

Some people often get dazed by the flash of imported building materials, but if you want to save money and reduce your carbon footprint, the best way to go is using local building materials. While it might always be an available option for everyone, definitely try to find local materials for your shipping container home. Some countries are rich in certain resources, while others might lack the same ones. So, dig a little, and if you find the building materials you need locally, get them. This will significantly lower your carbon footprint, not to mention save you shipping money.

Some people take things a step further and source materials that are not only local but also surplus. If you look around in your area, you will probably find excess building materials and discarded raw materials that you can use, thus minimizing the need for buying new products. These materials might have some defects, but with a little effort, they can fit into your design and save you a lot of money because they are surplus.

There are many other things you can do for your shipping container home to make sure it is sustainable and poses minimal threat to the environment. From using energy-efficient light bulbs to water-efficient showerheads and toilets, the choice is yours. While these ideas will save you money in the long run, it is helping protect and save the environment that is the ultimate goal.

SHIPPING CONTAINER HOMES

CHAPTER 15.

UTILITIES

Utilities are necessities that you cannot afford to do without on your building site. You don't just need them in your container when you must have finished building, they would also help you to make building easy. There are companies responsible for servicing utilities in different locations, you can do an online search to find the ones servicing your building location. Before striking a deal, ensure you understand the rates and terms of payment. Some utilities involve minimum monthly charges. These charges start to count after installation. So might consider waiting to commence building before installing these utilities.

Some of these utilities are regulated by the government which is much more affordable. Endeavor to do your research well before settling for a utility company.

Electricity

This is one of the most essential utilities that you cannot do without even at construction. Contact an electrical company to get familiar with the processes required in setting up an electricity meter. If there is a power line across your road, it is much easier to get power installed into your container building. The cost is most times dependent on if there will be the need to buy a transformer for your container, the difficulty of the installation process, the installation location (poles or underground).

The utility company usually provides a short length of poles and wire; if you want to use it for a longer distance, you'll need to pay for extra wire. You should get an estimate for the extra amount of wire you'll be using. You might get a discount on your first month's service fee. Another issue that you might have is that some electrical companies might desire that your building attains some form of progress before they install electrical appliances in it. They might have reservations about the possibility of you finding the building and maintaining a beneficial customer relationship with them. This is not so with every company, so do your research well to understand the company's policies before you sign up for anything. You might have to start your building on temporary power, at least it would provide you with short electrical circuits. You can use it for construction but it might be too small to power your appliances after building. After construction, you can ask for permanent service.

Gas

This is mostly used in space heating, for water heaters and stoves. It basically includes natural gas or propane. If you are building in the city, you can contact a gas distributor just the same way an electrical service agent works. If you are not close to the city, you can purchase a heavy tank that contains gas that can last for 1 month time or longer. However, endeavor to find the cost of gas in your location before going to make a decision.

Sewer and Septic

Different sewer lines have different policies and processes for tying in, you need to do your research to familiarize yourself with these processes before settling for an option. A septic line is best for rural areas. The setup is always costlier than a sewer system but after installation, it requires no cost for usage. Unlike the sewer system

UTILITIES

where you have to consistently pay to use. Septic tanks are usually buried with lines or pipes. You are to discuss and strategize with your contractor and installer on a location that won't disturb your construction process at the moment or in the future.

Telecommunication

If communication and connectivity are vital to you, there are various options you can consider to make sure you stay connected and have access to constant communication with your loved ones around the world.

There are several options available which spans from DSL, cable, fiber, and others that influence the television network services, Internet, and phone connectivity. In rural locations, you might need to use slower speed cable, satellite dishes, DSL connections, and point-to-point radiofrequency techniques. If you live in a location that allows you to have access to multiple options, do well to compare pricing rates, and availability to prevent haggling and poor network connectivity. You can make a better decision by consulting people living around there to know what they use and how they use it. Making sure you have active telecommunication services at an early point is highly necessary to aid google tracking of stores, and the installation of a security camera.

Water

Some developed states like the United States have provisions for neat water, their water is neat enough to serve you for bathing, laundry, and cooking. However, in some other underdeveloped locations, you might need to make separate provisions for your drinking water, probably need to get a portable plastic water can. Rural areas also have their provision for water supply, although it could be very stressful or difficult to access. You have the option of digging a well, a borehole or commissioning a water distribution company to fill your water tank periodically. Some of these options might have high upfront costs but is most profitable in the long run and is more hygienic.

SHIPPING CONTAINER HOMES

CHAPTER 16.

ROOF AND CEILING

Roof

Getting a roof for your container home is something that you could choose to do or not do, depending on your style and whether you have the money for it. However, when you roof your container, you end up saving yourself from several issues like energy bills and heat. Usually, with a roof, most of the heat that sits in your home could end up being lost by processes like convection.

Types of Roofing Styles

There are different kinds of roof styles that you could choose to use for your container, and they include the following;

Shed

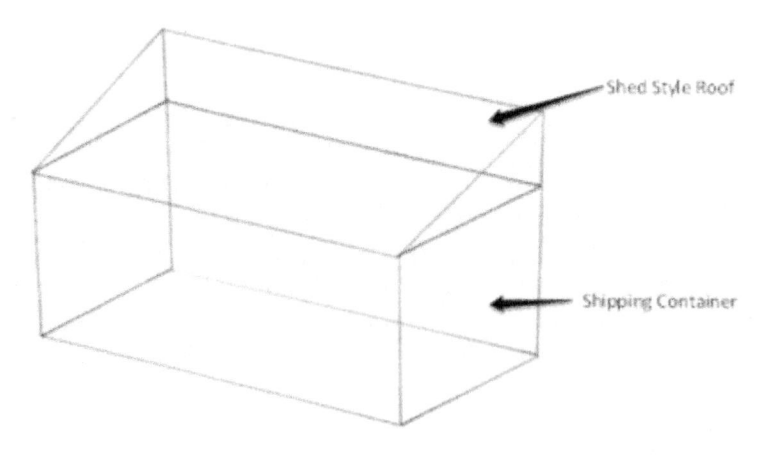

- This roof has a sloppy surface.
- It is very cheap to construct.
- It can be easily built.
- You can install this roof in a short period of time.
- The roof is one that works best if you plan on fixing solar panels for them.

To fix this roof, follow the instructions below:

- Weld right-angled plates across the length of the container roof.
- Fix a wooden beam onto steel plates.
- Screw-in the beams
- Use steel bars to support the structure.
- Use galvanized metal sheets to cover your roof.

Make sure that your roof can be easily ventilated. For this, ensure that the soffit board has a gap of about 1 inch in the middle of it. This gap can be covered with wire mesh.

ROOF AND CEILING

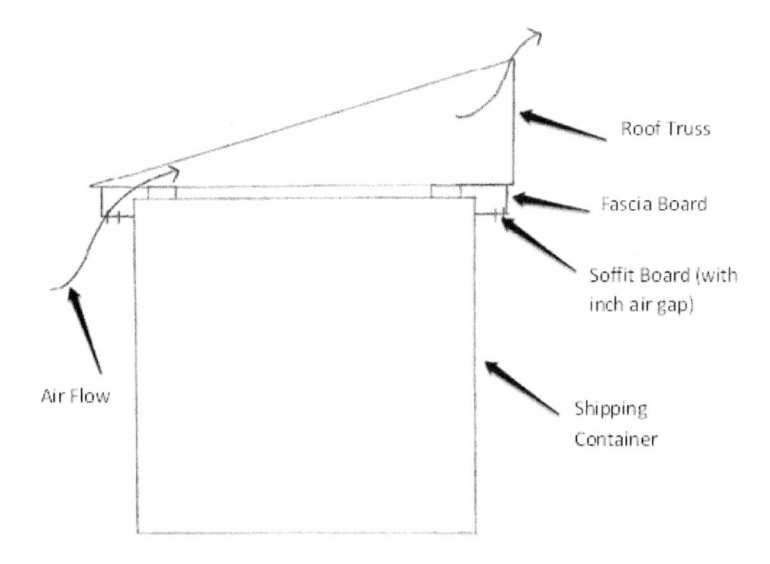

Gable

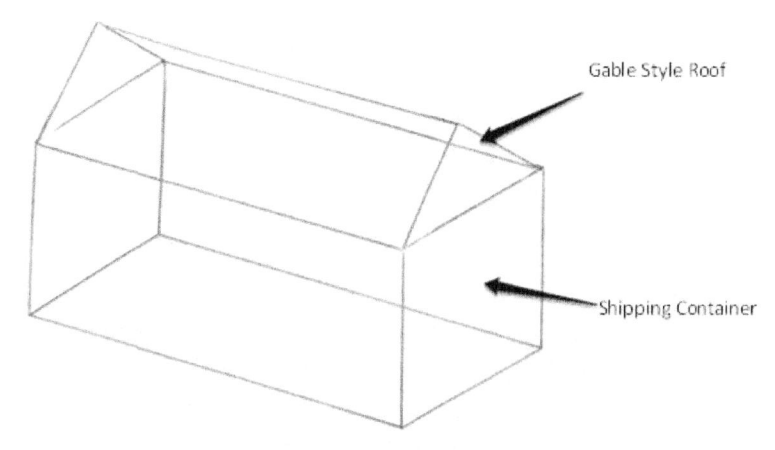

- ◆ It has a traditional and triangular outlook.
- ◆ It has a roof with a slope that helps to drain water easily.
- ◆ With this kind of roof, you are less likely to face leak issues.
- ◆ It offers more space for a ceiling than other kinds of roofs.

To install the roof, follow the instructions below:

- ◆ Weld a right-angled plate across the length of the container.
- ◆ Fix a wooden beam to the plates.
- ◆ Screw in the trusses to the plates.
- ◆ Fix the purlins across the trusses to finish the structure.

To ensure that the roof is well ventilated, ensure that the trusses overhang, just like it is shown below. You could also fix a fascia and a soffit board underneath these trusses.

SHIPPING CONTAINER HOMES

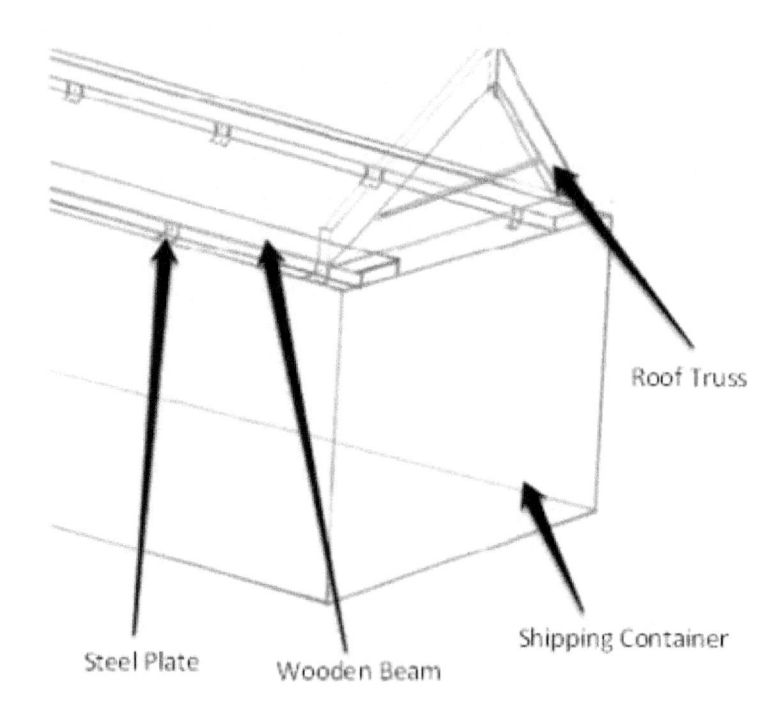

Flat

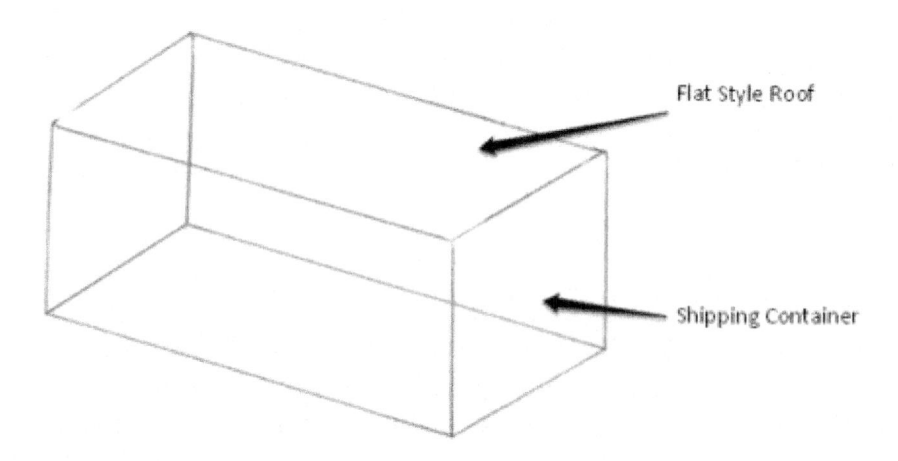

- This one is the roof the container originally has, and that could just be enough.
- The issue here is water stored up in the roof.
- You could lay a tarpaulin sheet on this kind of roof to protect it from rust and moisture.

Why You Need a Structural Engineer?

Having a structural engineer when building your roof will go a long way to help you because they would be able to help you make some necessary calculations like:

- **The dead load:** This is the weight of all the materials you use to build your roof.

ROOF AND CEILING

- **The live load:** This is the weight of the equipment and the people that install the roof.
- **The transient load:** This refers to a load of factors like rain, snow, and wind.

The amount of load your roof can carry is the amount of load it can bear before giving in. So, you must get someone who can help you make the right estimates so that you don't lose your roof.

Installing a Ceiling

Installing a ceiling, much like installing a roof, is a personal choice depending on your personal preferences. You don't have to put one in, but it will certainly make the place look a lot fancier and more to your style. However, an exposed ceiling also has its perks because it shows the nature of your shipping container home, so it might have an aesthetic appeal of its own. Still, adding a ceiling gives room for more insulation, and thus, better heat/cold control, and it might even provide you with storage space. Adding the ceiling before or after the stud walls is also a matter of choice, but you should do it after framing the walls. You have a few options with ceilings, starting with keeping the original one.

Exposed Ceiling

An exposed ceiling means leaving the container's original one, which goes to show the work that went into this home and making it what it is. It will also provide you with additional height since there aren't any additions. Exposed ceilings are a favorite for DIY enthusiasts since it entails saving a lot of time and money. But know that it means less insulation, so you might be paying that money and more in utility bills during winter and summer. Rust and mold might present as problems down the line due to condensation seeping to the roof from the house's interior, but it shouldn't be too serious, and you can fix such problems easily if they occur.

However, it is not recommended to skip out on the ceiling if you didn't add in an additional roof. Keeping both the original roof and original ceiling means you will have little insulation, if any, which can be quite problematic for a shipping container home. So, suppose you don't want to add a ceiling to the interior of the containers. There, it is definitely a good idea to add an additional roof on the outside so you can increase the insulation and dampen the effects of the sun hitting the roof all day long. The better option would certainly be to add a ceiling to the interior of your container. If that is not an option, then you should at least ensure that you have an additional roof.

If you don't want to leave an exposed ceiling, you need to install a new ceiling. Here's how you can start.

There are a few different ways to install a ceiling in your shipping container, but installing joists to the roof is the best way to go. Use 2.5 inches self-tapping screws to connect the joist directly to the roof of your container. Another way to go is to nail the joists into the head plates if you frame your house first. To save a few inches of space for insulation, you should start by attaching the joists directly to your container's roof beams, but this obviously needs to be done before framing the container and adding stud walls. You can then attach the head plates to the joists after framing the ceiling if you do this.

Make sure the joists are placed 1.5 inches between their centers, which will give you more than enough room for insulation. You can then add insulation between the joists to further improve the insulation of the container. After that, add panels or drywall over the joists and screw them in place.

FLOORING

CHAPTER 17.

FLOORING

Inspecting the Original Floors of the Container

The hardwoods in shipping containers are laminated from teak woods, and they are structured under marine requirements to make them durable and tough enough to withstand salty waters. Also, they are sterilized or treated with chemical treatment in accordance with the import/export regulations. Some of these floorings are known to inhabit viruses and contaminants. You need to ensure you discard them before beginning any work on your container interior. You would be able to escape this dilemma if you purchase new containers because it has not been used and it is void of contaminants. You could still indicate your preference for an alternative flooring type when ordering a used shipping container.

Considering the extra cost of flooring your container, you might decide to retain the original container floor. However, you have to ensure that it is safe and free of hazardous chemicals. You can check the chemical type used in flooring on the safe convention plate that is usually screwed to the front of the container. This plate contains 3 individual sections: the treatment date, chemical type, and immunity. You can discover how safe or hazardous it is from this information.

Here are things to check out when inspecting the original floors of the container. Usually, basic information is on the safe convention plate but there is always an exception of some categories of information. Some of the hidden information might include:

- Was there any damage to the flooring at any time?
- Was the damage replaced?
- What type of products was the container shipping?
- Did the chemical spill off on the floor in the process of transportation?

This information is majorly concealed by the supplier. Hence, you have to be sure that you're dealing with a trustworthy supplier, that's the only way you can figure out these answers.

However, you could still conduct your own individual check and validate the safety of your container floors.

If you are unfamiliar with the container safe convention plate information, you might not understand most of the information there when inspecting your container floors. On that plate, there is a section titled timber component treatment. It has the following details under it:

- Immunity (IM)
- Chemical used in treating the floor
- Date of chemical treatment

To identify harmful chemicals, check out the list provided by the WHO pesticide classification guide so you can understand the risk that each pesticide type includes. This plate is usually changed upon every treatment of the floor. So, you can trust that the information is updated and valid.

Asides from the safety of the flooring, you need to also check the structural accuracy and perfection of the container floors. All these decisions must be made before you start working on your container interiors because this is much more affordable.

Remove or Not Remove the Original Floor?

Although the cost could be a deterring factor in deciding what to do with your container floors, it is almost expected that you change the floors because a new floor is more durable and gives your building an aesthetic look. No matter how durable your container floor is you need to still consider removing it to place a new one. If you decide to still keep the original container floor, you need to epoxy it. Epoxy is a commercial way of sealing your container home floor. It will give it a whole new look entirely and increase its longevity. However, epoxy is quite toxic and would need to be handled with a respirator or full ventilation to reduce its toxicity. Do well to clean your plywood with a chemical preferably an isopropyl alcohol to give you the best bond.

Removing the Original Floor

Removing a container floor starts from cutting open holes around the place where there are bolts to easily remove it from the bolt. You can use a simple saw or a hand drill for a more perfect effect. After sawing, what is left is to pull out the wood. This could prove to be difficult in older shipping containers because over the years it has built tension and it is more difficult to pull out wood from the container. However, a medium-sized screw bar is a strong enough tool to put into work at this point.

If this proves to be quite stressful, consider cutting the wood into smaller bits and pulling every other part of the container floor that way. This is time-consuming but very effective.

Container Floor Replacement

If you have decided on removing and replacing the floors of your shipping container, there are other options you might like to consider that would give your container home a much homely feeling that you desire.

Shipping containers are cost-effective and convenient building alternatives. It is unique and conventional because the building space was formerly an industrial environment. Hence, they mostly come with hardwoods that are pumped up with harmful pesticides. The best alternative still remains to get an ideal floor type that is safe, suitable, and beautiful for your container home. This way you can comfortably enjoy the small space that is now a vibrant building you can live in without many hazards. If you would be living in the container, you would have to get rid of this floor for a better, safer one. There are various options open to you, but what is the best option for you? A few of these unique floor options have been listed for you.

Best Flooring for Flat Pack Shipping Container

- **Bamboo flooring:** This is the most affordable, flexible, and durable replacement option for flooring your container. There is no need to cut the bamboo

FLOORING

or remodify it into new shapes before fixing them into the container, the flooring dimensions are just the same as plywood floorings.

- **Carpet flooring:** If the container floor is not too toxic, you can retain it for use and look for an alternative for covering the surface so as to give your container a more aesthetic look. The carpet is a better covering option. This is the nice flexible and container flooring option; you can use different designs and colors in recreating a great look container home.
- **Steel flooring:** Most containers come with steel flooring and a nonslip feature. You place this steel on the original plywood flooring in the container. It is stronger, durable, and more water-resistant.
- **Vinyl flooring:** Vinyl is known to be a long-lasting flooring option; they commonly last for more than one decade before you could start noticing signs of wear. It is highly resistant to scratches, rusts, and stains. A good option, if you are intending to move in with children.
- **Linoleum flooring:** This final option is an eco-friendly and highly durable replacement option for flooring your container. With a natural antibacterial component, it has a natural biodegradable ability that enables it to last for 40 years and more.

SHIPPING CONTAINER HOMES

CHAPTER 18.

INTERIOR

Doors and Windows

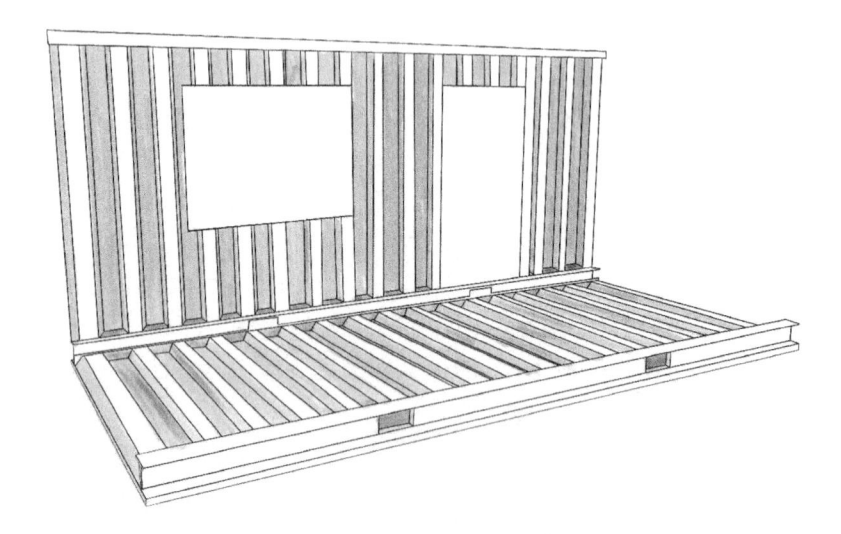

With the walls now open between the containers, your shipping container home should be taking shape quite nicely. Next, you need to start working on the doors and windows and the frames you need for each. The first step to working on windows and doors is taking their measurements and marking their locations on the walls. As always, these measurements need to be accurate and they have to be double-checked because you will be changing the frame of the container, so you can't afford to make any mistakes here. Some experts recommend using cardboard templates for all the doors and windows you will be working on and marking those. Get it right the first time.

Then, cut through the container walls following the measurements you've taken for the doors and windows. Use plasma torches, cutting wheels, and whatever tools you find necessary to do the job. Also, as with converting the walls, remember to wear protective gear because those cut parts will be sharp and can injure you if you are not careful. Make sure all rough edges are smoothed and fill any gaps in the metal walls with a sealant to ensure that your shipping container home is watertight and won't allow pests in. What comes next is creating the frames for the doors and windows, and after that, you install the doors and windows and hang them to the frames.

Making the Openings

To make the openings for walls and frames, the process is pretty much the same as cutting through the container walls to make more space. Like before, take accurate measurements of the required opening and mark it on the container wall. You can use a cardboard model of the window (don't forget to include the frame) which will

INTERIOR

help you get exact measurements. Then, cut through the walls with a torch or other tools.

A plasma cutter is probably the best tool to use here because it gives the cleanest lines and the steel you cut can be reused, unlike other tools that might damage that spare steel. If you can't get your hands on one, or you don't know how to use it, you can make it with an angle grinder which will do the job, though managing that tool is complicated because it doesn't easily make straight lines, so you must be patient. Last but not least, use a flap disk to smooth the edges and the opening.

How to Make the Frames

Before you can install the doors and windows, you need to make the frames, which will make your container pop and look like a real home. You can either order prefab frames designed for doors and windows, or you can make them. You could make a square out of galvanized steel tubes with 50 x 50 mm dimensions and cut several lengths of them for the frame. After that, put the frames against the doors and windows to make sure that the measurements are correct. If they fit, remove the doors and windows and stitch-weld the frame. For aesthetics and extra protection, smooth the constructed frame, and then spray-paint it with galvanized paint to resist corrosion.

Now, you have your openings and your frames, so you can weld the frames to the container after cleaning the edges and making sure they are smooth. After that, it is time to hang the doors and windows.

How to Install Doors and Windows

After welding the frames to the opening you have already made, is time for installing the doors and windows. Next, hang your walls and windows into the fixed frames and weld them together. You can also use self-tapping screws here to secure the windows and doors in place if you are not handy with welding, though welding is a more durable option and requires less maintenance in the long run. Remember to fill any gaps between the frames and the container with a sealant to maintain its structural integrity—pay special attention to the corners because they are the weakest part of the frame. You should repaint any metal parts with galvanized paint because they might become exposed during the installation process.

Installing Cabinets and Appliances

Cabinets and countertops give your kitchen area functionality. Even in a shipping container home, where space can be limited, it is important to carefully plan your kitchen cabinets and countertops so that your kitchen can be as functional and accommodating as possible. It doesn't make much sense to achieve the cost savings and simple way of living in a small size shipping container home if you have to spend lots of extra money eating out or relying on others to do things for you.

In a shipping container home, you may not have space for separate dining and living rooms. Therefore, a kitchen may be not only a place to cook food, but a family area, a social gathering area, and even a workspace. Consider all the things that are important to you, like hosting friends, baking, working from home or any other activities where you need to accommodate others and properly plan for them when building your kitchen.

Before diving into deciding which cabinets and counters you want, consider the plumbing and electrical system. It is very challenging in a shipping container home to run your wastewater lines. If you opted for a slab or raft foundation, you will want to remember to keep your kitchen sink and dishwasher bay as close to where your wastewater outlet pipe is so you don't have these pipes being run inside your container home, taking up valuable space.

In our original example of a 20-foot shipping container home, you will basically only have room for a kitchen, bathroom, and bedroom. If you are planning for a bigger container, you'll have more flexibility. A 20-foot container, however, will give you a very small home. Even with a larger container, you may still be fairly confined in how you plan. Planning the kitchen will take special care to accommodate all the needs you'll have as a homeowner.

Since we've oriented where the sink and dishwasher will be located—if you even decided to utilize a dishwasher—we must now consider where the kitchen sits in the house. If you have your exterior door entering into the kitchen, you must take into account the space you need for the door to swing. The door location and sink location should give you a good starting point on how to lay the rest of your kitchen out. Your next major pieces to put in place will be your refrigerator and range bays.

The biggest consideration with where your refrigerator and range will go is that you will have limited space above them for cabinets. It is best to have a range hood over your range to vent out the extra heat and smoke from cooking. Your shipping container home will fill up with smoke and heat very quickly. The space above a range hood may be utilized for a small spice or medicine cabinet. Above your refrigerator could be a small pantry area. Planning what to do with each little bit of space will let you eliminate any excess cabinetry that you don't need.

In order to determine the rest of your cabinet spaces, simply measure the space you have available and shop for what cabinets you feel match your desires and fit the spaces. It will be important to ensure that your cabinets are adequately secured to the wall when installed. Consider the weight of plates, glasses, and any other dishes or food items you may store in them. Use screws that will penetrate the majority of the stud size that you chose when building your walls.

If you opted to have the metal surface of the container exposed to the interior of your home, you may have to use metal screws or weld brackets onto your container in order to secure your cabinets. Metal screws are likely to be the best option. However, this is assuming you are putting some sort of exterior material around the

INTERIOR

container. Otherwise, you will create holes that can cause leaks. To effectively use metal screws, it will be best to drill pilot holes before driving the screws through. Some styles of self-tapping screws are available, but penetrating the thicker metal of a shipping container may be pushing beyond their abilities.

If you choose to use metal brackets to mount cabinets or any kind of fixture in your home, be sure to measure very carefully. If you measure just a little bit off, you can risk your cabinet or fixture not fitting properly. Cutting, grinding, and re-welding a mounting point can be very time-consuming. Also, be sure not to mount anything too soon after you made your welds, as the heat could damage the item. It would be wise to paint the brackets or mounting points once they cool as well in order to prevent rust.

When your cabinets are mounted, you may be faced with adding a countertop. Countertops can be purchased at most hardware stores and can be cut to various lengths depending on your needs. Some will have holes for the sinks pre-cut to fit a certain style of sink. If this works for you, great, but you can always cut the hole yourself for the exact sink you want.

Once the countertop is cut to the length you desire, installation is fairly simple. You may need some help to set the top down evenly. The easiest way to secure a countertop to a cabinet is to run small screws up through the frame of the base cabinet and into the bottom of the countertop. Obviously, you want to be very careful not to go too far with the screws, or they will puncture through the top of the countertop surface. Err on the side of caution and check how well it's holding by gently lifting up and shifting side to side. If you have any movement at all, try either adding screws or using slightly longer screws. If the base cabinets do not give good areas to screw through, utilize small metal brackets to better place your screws. To save money, you can even use small strips of scrap wood. This will all be hidden under your cabinets, so as long as it's secure, it's fine.

If they did not come with them already, be sure that your cabinets have all their necessary knobs and handles. If you have a missing outlet or switch covers, don't forget to put them on. Check what sizes of light bulbs you need if you haven't already. If you see space for shelving, consider what you could use it for and measure it out. Record all your small items on a list so you can knock them all out at once with one trip to the hardware store. They seem very minor, but for all the hard work you're doing, you want your house to look good and finished with every detail.

Adding Appliances and Fixtures

With your cabinets and countertops being set, you can go ahead and add in the appliances that really tie your kitchen together. This step will involve setting your refrigerator, range, microwave, dishwasher, and garbage disposal. Setting appliances is very simple in most cases, but it helps to keep a few things in mind. You will want to be careful when selecting appliances and think about how they will fit with the design you went in your kitchen. You don't want scenarios like having your refrigerator door open right up against a wall where it damages both the wall and refrigerator door. You don't want to fight with getting your new range oven into the house just to find out it doesn't fit between your cabinets.

When selecting a refrigerator, you would be wise to purchase the model you want at the same time you are planning what cabinets you want. These plans need to be in sync so that you don't have massive gaps between the top of your refrigerator and the overhead cabinets above. You also want to be mindful of how the doors to the refrigerator open. Some swing to the right, while others swing to the left, while still

SHIPPING CONTAINER HOMES

others open to both sides. Some freezers are on top, while some are on the bottom. If your refrigerator is against a wall or in the corner of an L-shaped cabinet layout, you may run into issues with where doors will swing. Lastly, you'll want to make sure your refrigerator is properly leveled. Most models will come with adjustable feet, but you can use shims if you find them to be significantly out of level.

Setting a range in your home is similar to setting your refrigerator. Your main concern with the range is to ensure you have properly wired an outlet for it to plug into. Many ranges will require a more powerful plug to operate. You also want to make sure you have some sort of fireproof under-panels under the cabinets that are adjacent to your oven. If you didn't do this step initially, you can easily add them in. Typically, the under-panels are made of thin drywall. Drywall does not burn; therefore, it will protect the bottom of your cabinets from catching fire if a fire started on top of your range. There should be adequate space recessed under the cabinet for the drywall sheets to occupy and not be seen.

Most ranges will require some sort of range hood over the top of them. The range hood has a fan that pulls hot air and smoke from cooking and vents it outside. While there are many styles of range hoods you can buy, to save space in your shipping container home, you will likely want to go with a microwave with a built-in range hood. This again should be planned before you start mounting cabinets in order to have the right cabinets to allow for a microwave. You also have to route the air duct up and out of the home. This duct will usually be concealed by the cabinets.

Microwaves can be somewhat tricky to install but the concept of installing them is simple. Most provide a bracket that gets screwed into the wall behind the microwave. This bracket should have several holes to hit the studs, no matter where they line up. Once the bracket is placed, drill small holes in the overhead cabinet to help support the microwave. Most microwaves come with a paper or cardboard template so you can accurately place your holes and brackets. You then simply hook the back of the microwave on the bracket and lift it up under the cabinets. As you hold it in place, you or a helper should run screws through the cabinet and into the top of the microwave. Once the threads are started, you can let go and tighten the screws until the microwave is flush with the cabinets and sitting level.

The duct work for the range hood typically won't connect directly to the microwave. Instead, the outlet on the microwave will align with the opening of the duct and they will butt together when the microwave is secured in place. If a particular model you purchase suggests something different, it is best to go with the manufacturer's recommendations.

When it comes time to set a dishwasher, double-check that you have properly allowed for an electrical hook up, a water hookup, and that your drain line is hooked up. The plumbing usually comes straight off the plumbing under the sink. Setting the dishwasher is not difficult, as long as you can connect all the necessary lines before setting it all the way back. It is best to test the dishwasher and have towels handy should something not quite be aligned right. Be sure the steam vent for the dishwasher is not directly underneath the edge of the cabinet, or, if it is unavoidable, consider adding a strip of laminate material to protect the wood in your countertop from the steam. Similar to the microwave installation, utilize the manufacturer's instructions to best ensure your dishwasher is properly installed. This will safeguard your warranty as well.

Another handy feature for a kitchen is a garbage disposal. Garbage disposals help shred and disintegrate food waste as it goes down the drain. They are, however, met with some mixed emotions from septic system professionals. If your home utilizes

INTERIOR

a septic system instead of being tied to a sewer line, you may want to be careful when considering if you want a garbage disposal system and how you'll utilize it. With a septic system, you want to avoid dumping solids down the drain as much as possible. A garbage disposal system is good in the way it breaks down the solids, but many tend to wash more solids down the drain as a result of the convenience of garbage disposal. If you are diligent in properly maintaining your septic system, you shouldn't have anything to worry about. However, for many, this often overlooked detail can be costly in septic system repairs.

If you decide you are the diligent type and want to continue with the installation of garbage disposal, you'll start by picking out the model that best meets your needs. This system will require an electrical hook up and even a switch that is easily accessible. Many switches will be located just inside a cabinet door or run up in the wall. A garbage disposal system doesn't technically require separate plumbing, but you will be modifying the existing plumbing from your sink. Garbage disposal is a somewhat complex component that will come with a warranty, so just as you should have with the microwave and dishwasher, follow the manufacturer's instructions for installation.

A feature in many new modern homes that continues to grow in popularity is the installation of water treatment systems. One of the most popular variations of water treatment solutions is the reverse osmosis system. These systems typically exist under your kitchen sink as well. If you desire this type of feature, you will have to carefully plan your space requirements to co-exist with the garbage disposal, if you choose to have one. In the case of a shipping container home, the space under the sink is also valuable storage space that may be eliminated with a reverse osmosis system. Though not as thorough in their filtration of water, many new refrigerators offer a filtered water spout. It all depends on what is important to you and how you want to live.

Regardless of whether you install a garbage disposal or dishwasher, the kitchen sink plumbing must be finished at some point. Plumbing the kitchen sink works similarly to plumbing the sink in the bathroom. You will install a P-trap to ensure the sewer or septic gasses do not make their way up the drain and into your home. The hot and cold water lines will use red and blue pex pipes. Faucets usually come in a kit with fairly self-explanatory instructions on how to install.

Installing Light Fixtures

Lighting is an extremely important factor in planning and building an effective kitchen. Your kitchen will be a place where you combine many activities that, in other types of homes, would have more designated spaces. Lighting underneath cabinets can have a big impact on how well-lit your kitchen area is. What makes wiring underneath cabinets convenient is that most of the work is easily hidden without the need to cover it up.

If you effectively planned out the other locations in your ceiling for lighting, installing light fixtures should be very simple. You can pick out a wide range of light fixtures at most hardware stores. Most will come with all the necessary pieces to install them. You simply need to make sure you have the wiring run to the right place and enough sticking out to work with.

For additional luxuries in your home, check out things like light dimming switches and various types of ceiling fans. Ceiling fans can dramatically improve the heating and cooling of your shipping container home. If you opted for simple window unit air conditioners and space heaters, ceiling fans can make a huge difference and leave you feeling very money-wise. Light dimming switches can allow you to install brighter lights for work or cooking but that is still comfortable for times of leisure or rest. If you are sharing a space for multiple activities like you would with a shipping container home, light dimming switches add more functionality to your home.

INTERIOR DESIGN IDEAS

CHAPTER 19.

INTERIOR DESIGN IDEAS

Designing the interior of your shipping container home should be similar to designing any other small home. You will want to fit in all your necessities and essentials without making your home seem crowded or, dare I say, tiny. Listed below are a few interior design ideas for shipping container homes so you can design a space-saving and aesthetically pleasing interior that suits your tastes.

Modern

A modern interior design style is sleek and uncluttered. It features wooden and earthy elements. The décor is usually minimalistic, and little to no patterns are used overall. If you want to give your home a modern and trendy feel, you can incorporate floor-to-ceiling windows and an open closet into your shipping container home. This helps to give it a more spacious feel. It is a great idea to use a color scheme consisting of brown, white, and grey hues, as it gives off a modern vibe without seeming cold and washed out. A color scheme with mostly white creates a simple and clean look while using dark wood tones adds a layer of warmth. You can also try to use modern, space-saving furniture like a wall-mounted TV and an all-in-one washer and dryer. It's also a great idea to use sleeping lofts, as it increases the amount of floor space you have.

Urban Chic

Similar to a modern interior design style, a chic urban shipping container home is classy and glamorous but not flashy. It is cosmopolitan and minimalistic, so it requires a lot of space-saving furniture. You should use plain walls combined with furniture in colors that are not "loud." For example, you can use black furniture with gray walls. You should also choose furniture with a straight-line design that gives off simple and stylish vibes.

The urban chic interior design style is perfect for shipping container homes as it makes use of small spaces. Some key architectural features include concrete, brick, and wood. The color palette is oftentimes neutral and the décor used is vintage.

Rustic

NTERIOR DESIGN IDEAS

You can opt for a rustic look to contrast the hardness and rigidity of your shipping container. Having a rustic theme will make your home seem more homely. Try using pine on your walls and ceiling; it borders between rustic and modern, so you will be able to have the best of both worlds. You can also use wood and stone as elements in the furniture, walls, ceilings, and floor. It will give your home an earthly feel. In the kitchen, you can try opting for a built-in counter that works as a table (for eating or otherwise) and as a storage space. Using this to save space can give you a chance to have full-sized appliances in your kitchen.

Farmhouse

This interior design style is perfect for sliding barn doors. It has an unmistakably relaxed vibe and is more homely compared to a rustic design. It contains both practicality and elegance, so it's perfect for working professionals who still know when to let loose and have fun. It does have some semblance to a rustic style as they both share a few elements. Modern farmhouses usually have weathered wood, brick walls or brick-like wallpaper, and exposed rustic wood beams. The walls, ceilings, floors, and even the furniture all have rustic finishes. You can try incorporating these into your own shipping container home if you want to have your very own modern farmhouse.

Colorful Accent

SHIPPING CONTAINER HOMES

The best thing about a shipping container home is the amount of creative freedom you have. Using dark stained wood combined with a bright color like teal can make your home look both stylish and practical. It is an unusual but aesthetically pleasing approach. You can try using bright colors on things like cabinetry or even a whole wall to create an accent and highlight the design and appeal of your home. Using cubby storage on the stairs can also be a great space-saving idea.

EXTERIOR

CHAPTER 20.

EXTERIOR

A lot of people choose to keep a shipping container's original design, both for stylistic and budget purposes. While you can choose to keep a shipping container's original exterior, you may also choose to personalize your home by using different materials that match the style you are going for. There are many options for the exterior of a shipping container home, and you can use the same materials you have been using to make the design of your home come together. If you need to add insulation to the exterior of your home due to space constraints, you can do so in a variety of methods with claddings or sidings. Adding external finishes to your home will assist in making it more durable by providing a protective covering. Aside from that, you can design a unique and personalized shipping container home.

The Style of Your Shipping Container Home

There are numerous options available when it comes to selecting the style of your home. Although it is recommended that you use environmentally friendly solutions because they are better for both you and the environment, you may also utilize other materials available on the market.

- ◆Modern. Steel siding, vinyl siding, or metal siding can all be used to give your home a modern appeal. Using these helps to imitate the appearance of numerous commercial structures.
- ◆Rustic. A wood covering can give your shipping container a cabin-like atmosphere for a rustic design. Log siding will have the same appearance as wood cladding. Pine and cedar are also excellent choices for this style of finish.
- ◆Lap and smart siding. Try lap or smart siding if you want a modern style but do not want your container to look like it's made of shiny steel. This siding is comprised of engineered wood that has been treated for further durability.
- ◆Combination. You may also use a combination of all the styles to create a unique design for your home.

How to Install Cladding Onto Shipping Container Homes

Cladding is the process of covering a building's external structure with a single material. It can wrap around an entire structure or be utilized solely on certain areas of it. Cladding serves to protect a building's structure from natural elements such as wind and rain, but it can also provide insulation, noise reduction, and improve the visual appeal of a house.

Installing cladding, when built with installation, on the outside of container walls is more difficult than it is for traditional wooden-frame buildings. The lack of pressboard supporting installation with nails or staples is what makes a huge difference, and therefore, complicates the process. Using metal screws is not only inconvenient and time-consuming, but it may also jeopardize the quality of your home.

SHIPPING CONTAINER HOMES

Cladding Only

Adding external cladding for aesthetic purposes is simply easier. It is because there is no risk of air intrusion between the structural walls and the cladding. Allowing a small gap between the container's walls and the outer cladding will allow air to circulate by convection. It is worth noting that air sucked from the bottom level will most likely be cooler than air sucked from the higher level of the walls. This natural passive ventilation will aid in reducing the effects of hyperthermia caused by direct sun exposure. You will need a supporting frame, but it can be constructed so that it is just loosely tied to the container's construction. The frame can be adhered to the container construction with adhesives, if necessary, but bolting it to the container structure is not recommended.

Cladding Over Exterior Insulation

Adding external insulation complicates matters. You do not want any air to get under the sheathing, especially between the insulation layer and the container walls. When you expose insulation to moisture, you risk it deteriorating. Outside air infiltration may also form "heat bridges," negating the benefits of the outer thermal insulation. It is advised that you use closed-cell spray foam insulations. These attach securely to metal walls and are virtually impermeable, despite their high cost. They may still grow mold/mildew if exposed to moisture. Hence, it may be necessary to build breathable cladding to allow moisture and water vapor to escape.

How to Install Siding Onto Shipping Container Homes

Sidings and claddings are similar in appearance and purpose, although siding is more commonly used to describe vertically or horizontally installed wood, vinyl, aluminum, or engineered materials, whilst cladding is more commonly used to describe brick, stucco, stone, or similar materials.

Special procedures are utilized to connect blocks to the metal walls of shipping containers without producing holes that could damage them. Installing siding on your shipping container can be done in one of these ways:

- **Wooden blocking:** To use wooden blocking in installing sidings, split 2×6 pressure-treated timber planks must be glued to the flutes of the shipping container. The flutes are used as blocking in this manner. In terms of construction, a flute is an inside and outward profile of a surface.
- **Heavy gauge metal blocking:** To use heavy gauge metal blocking in installing sidings, thick gauge metal "hat" furring channels will be utilized. Before applying the siding, be sure to paint over the welded area with rust-inhibiting paint.
- **Combination of metal angles and wooden blocking:** Using a combination of metal angles and wooden blocking is one of the fastest and chunkiest methods of the 3. The siding is secured on pressure-treated woodblocks. The wood blocking is fastened to painted metal angles in a continuous pattern, and the painted metal angles are welded to the shipping container's outside. Just like in the previous method, it is important to coat both sides of the welded surfaces with rust-inhibiting paint before covering it with siding.

EXTERIOR DESIGN IDEAS

CHAPTER 21.

EXTERIOR DESIGN IDEAS

Cladding Materials

Cladding materials are very many out there, but you will only find a few of them that can be used for containers. Since the walls of a container home are not breathable, you have to ensure that you use outer cladding to allow the influx of air. This technique is so that the moisture that peradventure gets in through the pores can be allowed the freedom to leave. If the moisture remains trapped, usually, what happens is that the structure gets destroyed due to the build-up of moisture. It could also cause wooden supports to undergo rots. The following characteristics determine the kind of cladding material you go for:

- The color of your container house
- The type of finish you employ for your container house.
- The cost of the materials that you employ for the construction of your container house.
- The R-factor.

Timber

Timber is one material from which you can obtain wooden logs. When these wooden logs are used in beautifying your container homes, you get something that looks as striking as anything. It is one material that gives the medieval feel, and it only gets better when you coat the surface of the timber with natural colors. The most commonly used wood includes the Canadian Western Red Cedar and the Redwood. The other sets of hardwood include chestnut and oakwood. Even without them being treated, these wood types can still come out as one of the best by resisting moisture and the effect of decaying organisms. You could also get them in colors like gold, red, and grey.

Timber exists in different forms, and they include clapboards, shingles, and half-logs which you can utilize when employing the overlapping and joining technique. To fix timber logs in your container home, you need not have too much technical know-how. Just by selectively interlocking wooden pieces for aesthetic purposes, you can easily get what you want.

SHIPPING CONTAINER HOMES

Bamboo

Bamboo is one tree that grows very quickly, so, you using them to decorate the exterior of your container home could come out as one of the best decisions you could ever make. They are usually used to decorate your wood as they exist in the form of tubular stems. When they are exposed to high humidity, you usually will find out that they can stand the test of time. The only issue you could experience with them is that they could lose their natural hues and have mildew growing across their lengths. So, to prevent the time the fading actually occurs, the stems are usually coated. Bamboo stems are usually cut open along their lengths, after which the outermost layers are peeled off to ensure that their natural hue is kept.

Engineered Wood

This wood merges the natural outlook of natural wood and the durability of synthetic materials. They are also known as wood-plastic composites. They are composed of about 50% of powdered wood fibers and about 45% of high-density polyethylene. High-density polyethylene is a non-toxic constituent that is employed in the food industry. They aren't receptive to factors like mildew, mold, or rot. They also offer a lot of resistance to ultraviolet rays and thermal effects. This wood exists in various colors, and it usually is all about choosing whichever suits you. Engineered wood, unfortunately, cannot be stained, so the moment you fix them, you can't really change its colors.

EXTERIOR DESIGN IDEAS

Fiber-Cement Boards

These boards are made from a mix of concrete and some other substances like fillers and pigments. Usually, concrete is found in higher percentages in the boards, and the fibers are used to support the concrete by a factor of about 10%. You could also add concrete circles of alumina that will help fetch you a final product that is rigid and unchanging. Since fiber-cement boards are constructed from natural sources, you'd find out that they are resistant to ultraviolet rays, rots, and fire. They can also be found in several shapes and colors. These boards usually are available colored, but then, you could repaint them if you desire.

Composite Materials

These are mixtures of 2 or more materials that help to provide a product that is way better than each of the other constituents. There are several reasons why you'd want to try out composite materials for the outdoors, and they include the following characteristics:

- Strength
- Lightweight
- Moisture-resistance
- UV-ray resistance

SHIPPING CONTAINER HOMES

- Low-temperature resistance
- Resistance to insect attack
- Resistance to discoloring

Even though they may be purchased at a fairly high price, composite materials come out as very elegant and durable. They are also non-biodegradable. Engineered wood is usually seen as one of the many composite materials available.

Composite Panels

These panels exist as layers of insulation between metallic sheets. The sheets usually are made of aluminum or steel. This feature gives them the name, "sandwich." They come out as a principal component of cladding and help to insulate your container home. If you plan to construct walls that will bear a lot of weight, it is best to use composite panels for the job. The composite panels have folded metallic sheets on the outer surface. They have a metal finish and are resistant to factors like moisture, mold, mildew, insects, and rots. Their initial colors usually do not fade with time.

EXTERIOR DESIGN IDEAS

Metal

Fixing a sheet of metal onto another sheet may not make too much of a sense, but then, it offers a lot of good treats. It is not affected by insects, ultraviolet rays, or even fire. You can clean the surfaces or have them repainted if some color fading occurs. Metallic claddings are made from metals like aluminum, copper, brass, stainless steel, and so on. Those made of copper alloys are usually more costly.

SHIPPING CONTAINER HOMES

CHAPTER 22.

HOW TO PAINT THE CONTAINER

Painting your container will help you beautify your house even more as it adds a luster to the original mundane color of your container home. When a professional oversees the painting process, they will ensure the paint job comes out great and lasts a long time. Professionals know how to prep the container to get a smooth surface they can paint on, and they are also knowledgeable on the right products to use.

The Kind of Paint to Be Used

For the best possible outcome, containers should be painted with primer paint and exterior premium paint in one—both of which come in satin or flat finish. On the other hand, a separate primer can be applied and then a topcoat. Two coats of paint are most times enough to ensure it lasts, but you may require more depending on your choice of color and the needed coverage. Also, you can apply insulation coatings on the top paint to prevent and deflect heat and anti-corrosion coatings; this helps to add strength to the rusted or corroded areas of the container.

Several colors of paints are available that can be chosen for your container, but then note that when making your selection:

- Lighter colors will need extra coats of paint for the right coverage to be gotten. Light colors will also display marks or dirt much easier.
- Darker colors may cause the container's interior to be hotter in summer, but they better mask imperfections than lighter colors.
- Use colors that blend nicely with your ambient environment.

When you use the right kind of paint, the lifespan of your shipping container can be prolonged and made to look brand new.

GENERAL ADVICE AND HOW TO SAVE MONEY

CHAPTER 23.

GENERAL ADVICE AND HOW TO SAVE MONEY

It is easy to get caught up in all the hype about shipping container homes. We can all appreciate the value of repurposing these hefty boxes to save the environment and no one can deny how innovative and creative they can be. But probably the biggest reason people look to build a shipping container home is the savings. There is no question that with the right planning you can literally save thousands of dollars on your home compared to the total cost of building a traditional standing home.

However, this cost savings has turned into a major case of sticker shock for many. While it is possible to save big on these innovative home designs, it is not always the case. If your goal is to beat the economy by saving big on a shipping container home, there are a lot of truths you'll have to come to understand. You'll need to become a very savvy investor and learn everything you can about the financial traps you can easily fall into. That way, you will know exactly what things you can cut out and what things you can't afford to skimp on. It will do you no good to save big on your housing project only to develop something that will be impossible to live in.

Careful Planning

Your plan should be laid out completely from the beginning. It is the only way you'll be able to gauge your success throughout the project. In order to save on the building costs, you don't want to find yourself improvising here and there in hopes of getting a good result. It not only won't lead you to a money-saving, budget-friendly home but is a sure-fire way to create major problems that will suck more money out of your budget along the way.

The first phase of your planning will be to decide exactly how much money you plan to spend on your new home. This will make it clear to you just how elaborate or extensive your building project will be. Once you have your budget completely figured out then it will be easy to determine exactly what features you want to have in your new home. Start with budgeting for the foundation, water, sewage, and power, and a basic structure. These are the essentials no home can be without.

While beginners and amateurs are totally capable of building a shipping container home themselves, there are major cost-saving advantages to hiring an architect or a contractor to assist you in laying out the plans. Yes, it will be an added expense but it will save you from making costly mistakes along the way. One of the biggest mistakes many make with these projects is removing too much steel for windows, doorways, skylights, etc. Removing too much steel compromises the integrity of the structure and could put your whole house in jeopardy. By using a contractor to help with the design phase of the project you can avoid this costly mistake.

Decide where your money is going to be spent. There are 2 ways you can go in this regard, for cheap or for quality. It may seem strange to suggest but purchasing higher quality and more expensive containers can save you money in the long run. The areas where you can afford to go cheaper would be on things that can be replaced once the project is completed—light fixtures, appliances, paint, and flooring are some possibilities.

Another way to save on expenses is to purchase materials from salvage yards whenever possible. When you're ready to purchase flooring, windows, fixtures, appliances, or anything else to complete your house, check out demolition sites, junkyards, or recycling centers. You'll find that many supplies can be found at a low cost and sometimes even for free as long as you provide for the delivery yourself.

Your largest expenses in the home and those that you shouldn't try to skimp on are the cost of the trailer, the land, insulation, and architect and engineering fees. These will usually be fixed costs that you will have to pay upfront. Other fees are flexible and there are always money-saving alternatives you can choose from.

Creating a shipping container home can be a great way to save money but it should not be considered a given. The cost of your new home will depend on many factors. Taking the smart road and planning out the details and budget ahead of time can make a major difference in how much you save on the overall cost of the project.

EXAMPLE CASE STUDIES

EXTRA

EXAMPLE CASE STUDIES

The cost of the entire build varies from project to project, you might have a very small budget to work with and the versatility of a shipping container home will allow you to get a very nice finished product for your money. There is not such a thing as cheap construction and, if there is, then you should be very wary of it, generally cheap construction offers cut corners on safety, permits, and the durability of the materials used.

This is not ideal when building a home you will want to live in for a long time. The shipping container itself might be quite sturdy but if you use less than optimal materials then all that potential will be wasted and there is a chance that you might end up spending even more money on repairing costs and maintenance. It is perfectly possible to have a cheap, sturdy, and durable shipping container home, it is only a matter of spending your resources well.

This is how you build a shipping container home with a small budget. Cheap does not necessarily mean ugly; however, when you have a small budget then lowering your expectations of how the completed house will look is good advice, forget about those luxury, multilevel homes and focus on getting what you need. It is possible to build a shipping container home for less than $30,000, to manage this, here are some things to take into account:

- As with any shipping container home, the planning stage is vital; for a cheap shipping container home, it is critical that you are as thorough as possible. This means that you need to have a clear building schedule with realistic expectations before you start the actual build, improvising will end up costing you an inordinately amount of money without you getting any benefit from it.
- Plan your budget, decide exactly how much you want to spend on your home and then add a little bit more for wriggle room. When you have established that, then you can move on to what it is what you want to build, how many rooms, and whether or not you will use a new shipping container.
- Wasting materials is the first money-draining cause in any building site and to avoid this you will have to be very clear on what you want your home to look like.
- Another point to consider when building on a small budget is what the cheapest containers in the market are and whether or not they are safe to use as the base for your construction. The shipping container accounts for the majority of the money spent while building the house, so this is something you need to have floating around in your head while you are planning. Buying used or new containers is the first choice you need to make, while buying used ones is significantly cheaper, you should only buy used shipping containers if you can take a look at them beforehand, otherwise, you might get a structurally unsound one and spend thousands on making it safe. Another tip is that you

make sure that all your containers are the same brand if you are using more than one.

- ◆ The most cost-effective shipping container out there are one-trip containers. These are containers that have only been used once for haulage and there is actually a very large number of them because China is mass producing them so take a look at some of those.
- ◆ Another great way of making your shipping container home quite cheap is to try and salvage local materials. This is not to say that you should take all the used materials you can and use them to build your home but it does mean that you should try to see what your community offers before outsourcing. Try demolition sites! Most of the time they will give out material for free if you pay for the transportation.
- ◆ Building a shipping container home is exactly like building anything else. Three things will impact the cost of the build: time, cost, and quality. Where it is quite hard to reconcile the three of them, meaning that you can build a shipping container home really quickly and to a high-quality standard but this will make the cost of it skyrocket, or you can build an excellent quality home for a small amount of money but it will definitely take more time. To make the building process easier you will need to decide which factors are fixed, meaning that you will have to set your priorities from the very beginning and make them clear to your contractor before you start to build.

Now, let's break down the main source of expenses in a build and expand on them a little:

An important thing to take into consideration is the insulation of your shipping container home. One of the worst mistakes you can make is trying to cut costs in insulation; if you get it wrong you will only succeed in making your home inhospitable. The most advisable way to insulate your home is to use spray foam insulation, it is the only one that can ensure a seamless vapor barrier that will help you prevent mold and dampness, it also happens to be the thinnest at 2 inches thick. The cost is a little higher than that of other insulation methods at $1.75 to $3 per square foot.

The second type of insulation is panel insulation. Panel insulation is cheaper than spray foam insulation and is slightly thicker than it, at around 3 inches of thickness. The biggest disadvantage of using panel insulation is that it needs wooden or metal studs to be fitted into the container which will reduce the internal space of your home by 90 millimeters all around. The estimated cost is anything from $0.75 to $1.45 per square foot.

Now we are going to look at foundation costs, this is yet another aspect where you should avoid cutting costs. We have already mentioned concrete piers as a method of foundation and highlighted its advantage as a DIY-friendly way to cement your home. Each concrete block is 50 centimeters x 50 centimeters x 50 centimeters—for a 20-foot container you will need 4 of these and for a 40-foot container you will need 6. The estimated cost for each block is $90, which is $550 for a 40-foot container.

EXAMPLE CASE STUDIES

A type of foundation that we did not mention is the strip or trench foundation method. This method consists of laying a small strip of concrete, generally 2 feet wide and 4 feet deep, around your container. This sort of foundation is perfect for softer soil that will not allow for simpler, concrete pier foundation types. This kind of foundation is a little more expensive as you will have to excavate a little bit more and will need more concrete. The estimated cost for a 40-foot container is around $5,400.

We mentioned the slab foundation type, now we are going to break down the cost of it for you. As mentioned, it involves laying a 10 to 14-inch-deep concrete slab under the container. Since the slab goes under the entire container, this method involves a lot more excavating than with previous methods. This method is best used in very soft soil that does not allow for a pier or strip foundation type and the estimated cost of a slab foundation for a 40-foot container will be around $5,900.

The last thing you should look into is the external cladding, this is the finishing of your home. It is quite popular to leave the containers bare for that "industrial look" that modern architects love so much. This is definitely the cheapest option as you will only need to buy a suitable type of paint. Unfortunately, leaving the containers bare is not always an option due to zoning restrictions that will ask that you make your container home blend in with the rest of the neighborhood homes.

The first cladding material to be discussed is called stucco (a thicker cement sheet with a pattern), which is actually a very popular cladding material and is widely used in regular homes. This will allow your home to blend in more easily and will also give your container another layer of weather protection. Applying a coarsely mixed layer of stucco directly onto your container has an estimated cost of $6 to $10 per square foot.

Another popular cladding material is timber which is great for providing a "natural look" to your container home, thus making it a perfect finishing technic for homes in rural areas. Despite the fact that timber can be quite expensive, especially external timber, you can finish your house cheaply by finding recycled timber. The finishing technique consists on fitting vertical battens to the outside perimeter of your container and then fixing the timber cladding to them. You may use bolts to fix the battens to the shipping container and nails to fit the timber cladding to the battens. The best choice for cladding material is Western Red Cedar but the choice is up to you. The estimated cost is $4 to $5 per square foot.

We will go through one complete material guideline build with you, to allow you to see what type of quotes are needed in order to build a shipping container home:

Your potential shipping container home:

SHIPPING CONTAINER HOMES

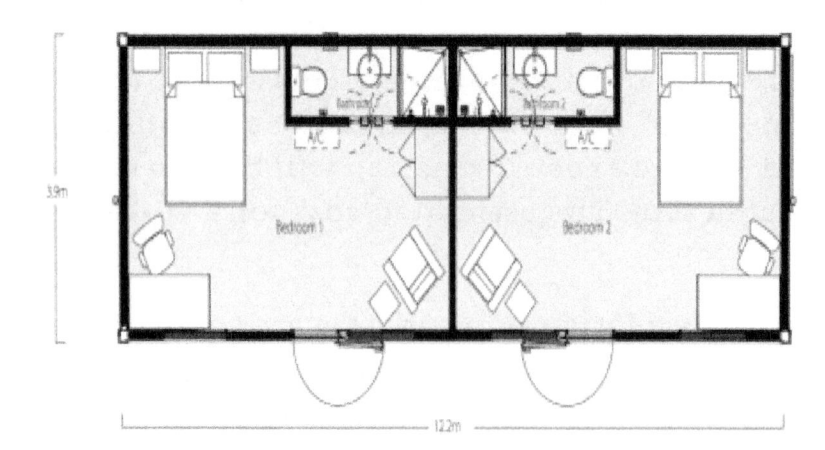

- Shipping container: We have 2 shipping containers here and we will use 2 40-foot high cube shipping containers: Used $2,950 a total of $5,900 then shipping costs are a big variable but you can use the company below if you wish. http://www.onetripcontainer.com/one-trip-containers/40-one-trip-container/40-high-cube-one-trip-container/
- Shipping costs to America from a nearby provider: $780 to transport a larger 40-foot container. This figure included the surcharge of unloading the container and the total journey distance was around 300 miles.
- The total cost of our 2 containers including shipping is $7,460.
- For painting the outside and inside we will need to use an undercoat, normal coat, and top coat: $300 including brushes/rollers.
- For 2 large windows of aluminum and double glazed of 1,500 millimeters wide by 2,100 millimeters tall: $760.
- Hume doors, a 2,040 by 820 timber front door: $179 each x 2 = $ 358
- Double sliding bifold doors: $220 each x 4 = $880
- Labor for installing windows and doors: $2000 for the lot.
- Timber flooring: Western Jarrah laminate flooring $54 for 2 square meters. We have here 48 square meters, we need all of the bearers underneath and nails/screws which are an extra $500. So, the total cost for materials of the flooring is $1,800 with labor is $4,000.
- For the electrical work, we want 4 light switches and 6 power points. All the wiring and material costs: $900.
- For the plumbing work, we will be looking at a rough-in and a fit off of 2 bathrooms: $960.
- To build walls by a carpenter and then plastered by a plasterer: 3,000 for all of them, including insulation around the whole building.
- Mondella 2,000 x 1,180 x 880 millimeters Concerto Shower Screen x 2: $900.
- Shower bases with a center draining hole x 2: $380.
- Shower head = $75 x 2: $150.
- Vanity unit 900 millimeters café oak x 2: $1,340.

EXAMPLE CASE STUDIES

- For the toilet x 2: $360.
- Air conditioner evaporative cooler x 2: $4,000.
- Kitchen bench with stove and overhead cupboard x 2: $1,800.
- Tiles for the bathrooms: $47 for a 10 pack—and we need 4 of those packs so that will be $188 and for grout and labor add another $1,000.
- For 2 queen-sized beds and mattresses: $3,600.
- Bedside tables: $900 for 4.
- A computer desk and chair in each room: $700.
- For 2 single arm-chair reclining chairs: $2,700
- The total cost for materials of a basic house plan: $38,636.

This example quote is perfect to help you understand how and what you need to do to budget for your home.

If you do this yourself and go purchase all the required materials, then you will save a lot more money. As I am in the building industry and I know how much all of the materials are getting marked up, you will save at least 15% on your build just through material mark up.

To put that into perspective for you, out of that $38,636 you can save a huge $3,860—so, if you can even get assistance from a builder/carpenter friend in what would be required for your home, you should definitely do it.

SHIPPING CONTAINER HOMES

FULL-SCALE FLOOR PLAN IDEAS

These are just a few of the designs we could show you to showcase the potential for customization in shipping container homes. Have a look and let the inspiration strike!

Plan 1

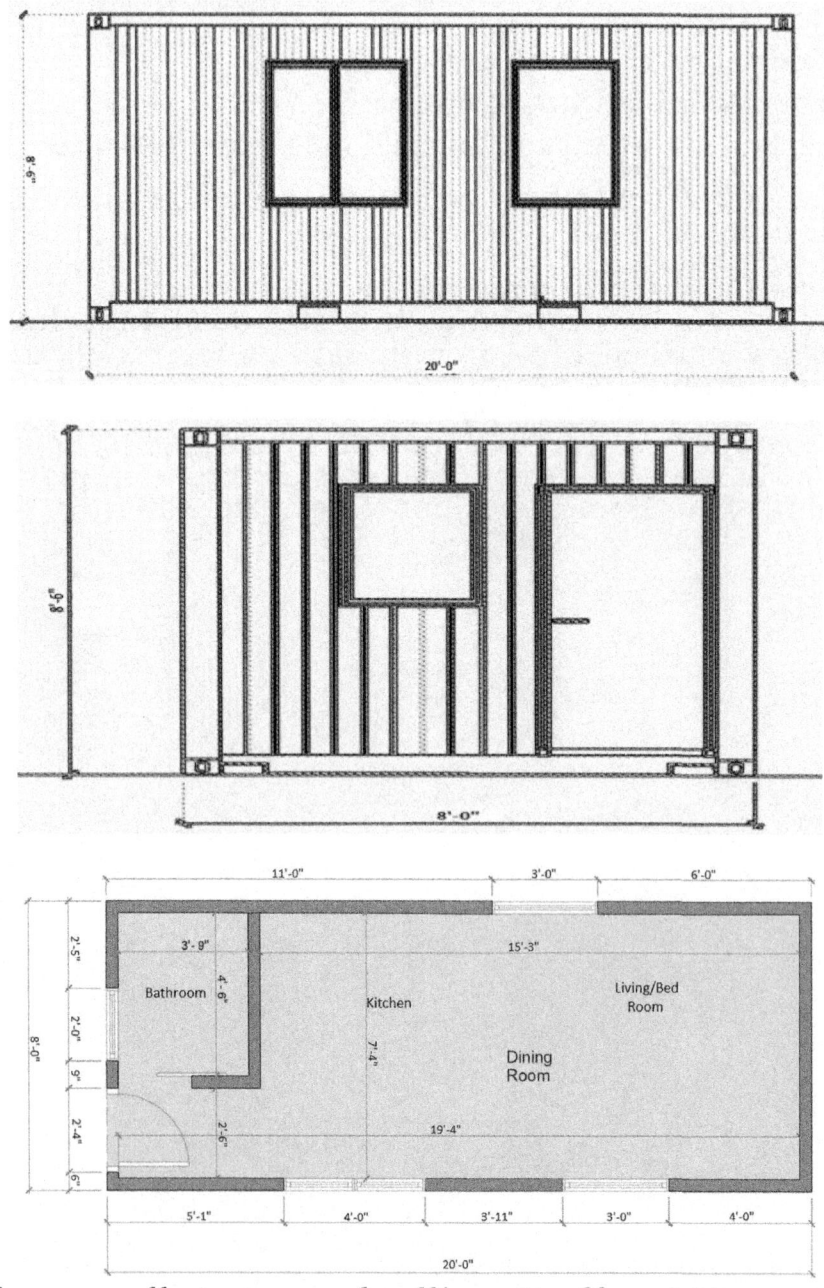

Plan 1 is ideal for a small, 1-person dwelling. It offers 139 square feet and is designed from a single 20-foot container and features a combination kitchen, dining room, living room, and bedroom. The ultimate in space efficiency and tiny living space.

FULL-SCALE FLOOR PLAN IDEAS

Plan 2

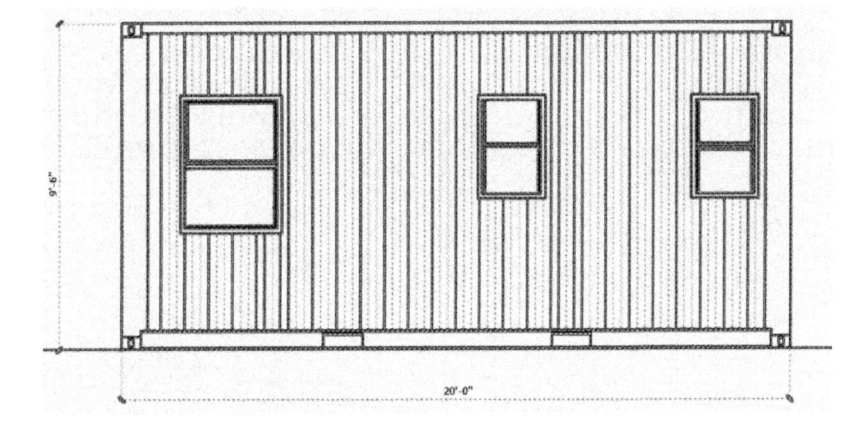

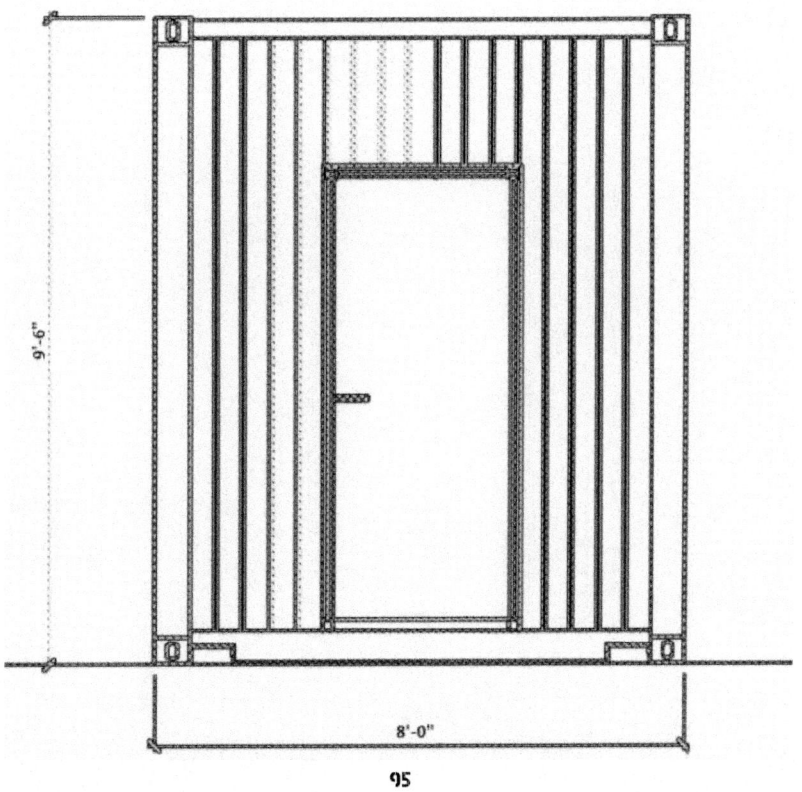

95

SHIPPING CONTAINER HOMES

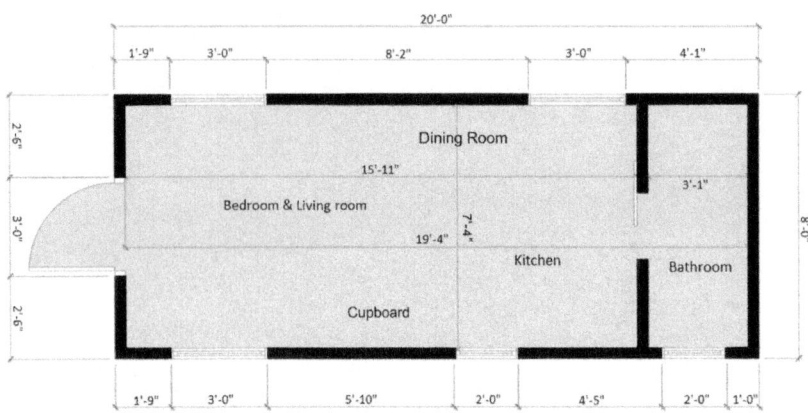

Plan 2 is another example of a single 20-foot container dwelling. It has been customized to offer a spacious bathroom and kitchen. The living room doubles as a bedroom and features 2 pull-out double beds, offering comfort whether sleeping or sitting. This economical design can comfortably house 2 people within 138 square feet.

FULL-SCALE FLOOR PLAN IDEAS

Plan 3

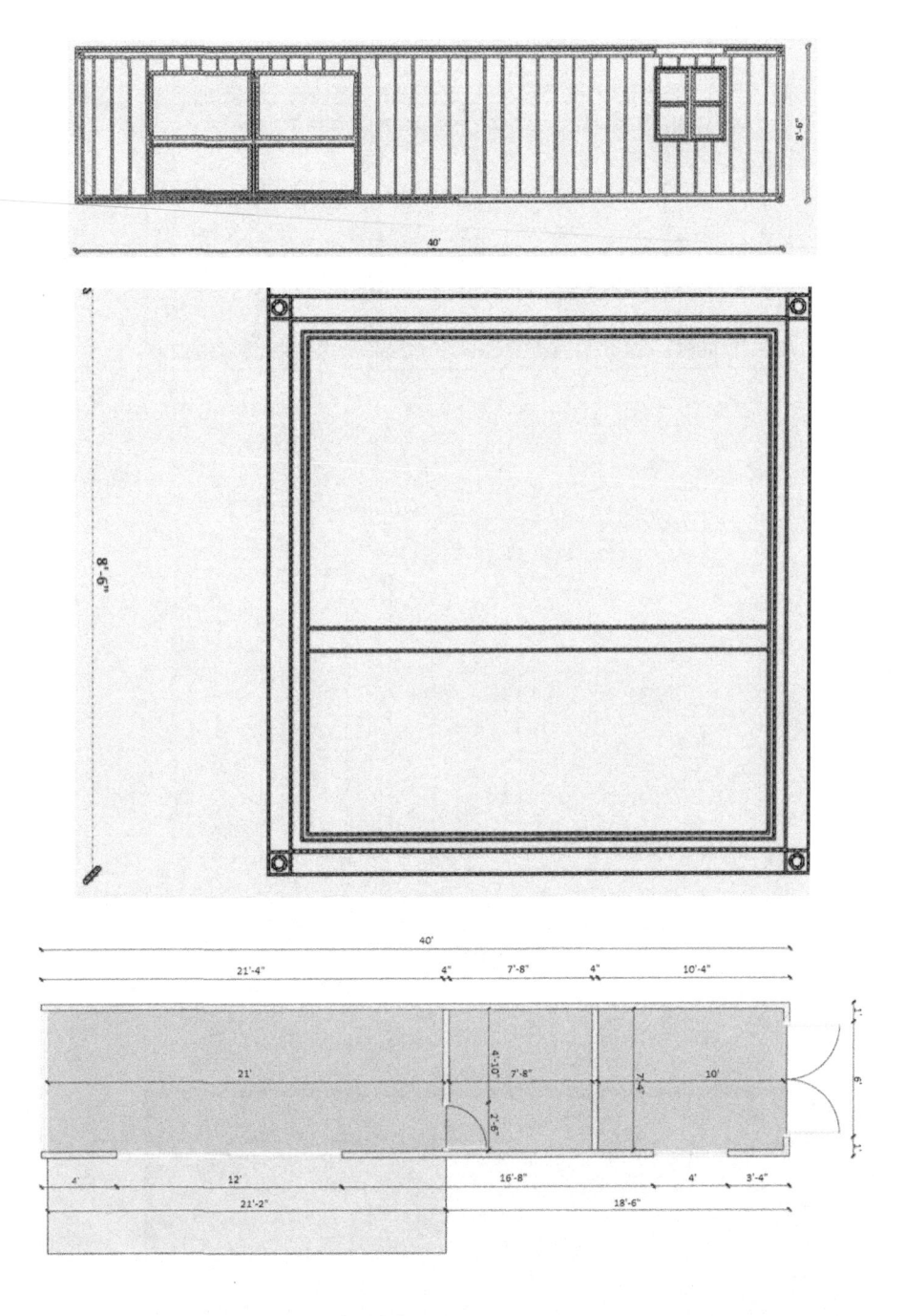

Plan 3 is designed from a single 40-foot container and boasts a spacious open-plan living room which would be ideal for a sofa bed. It features sliding glass doors which lead to an open deck. Also, included is a second room ideal either for storage space or for a second bedroom.

SHIPPING CONTAINER HOMES

Plan 4

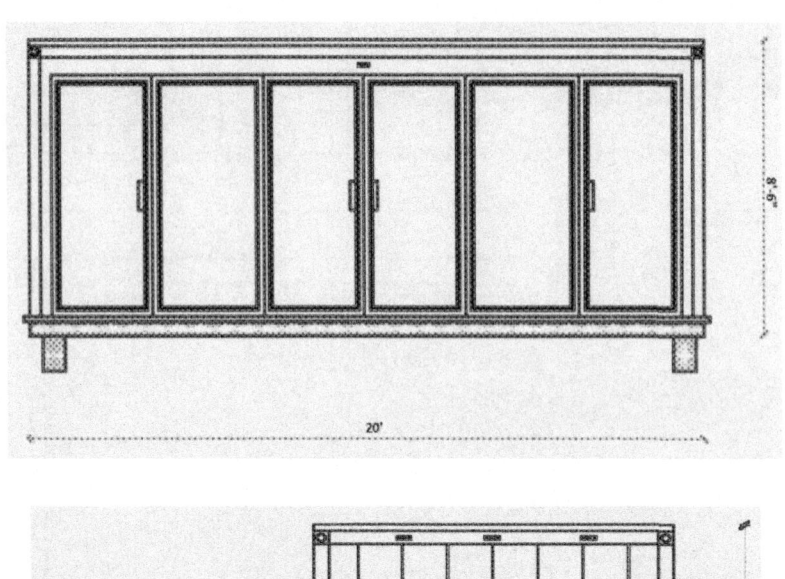

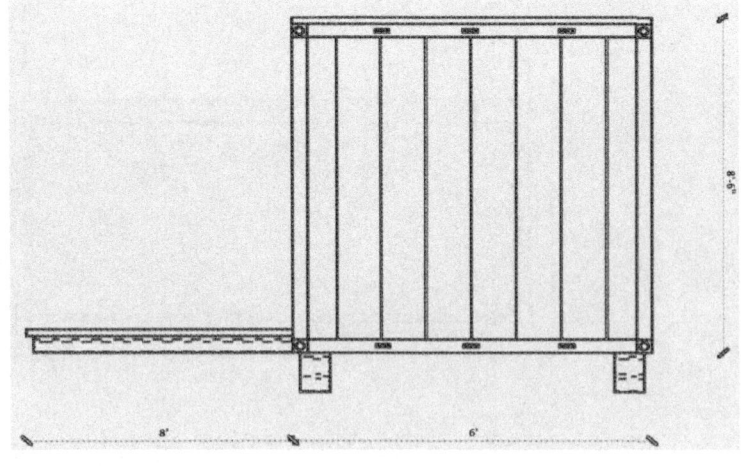

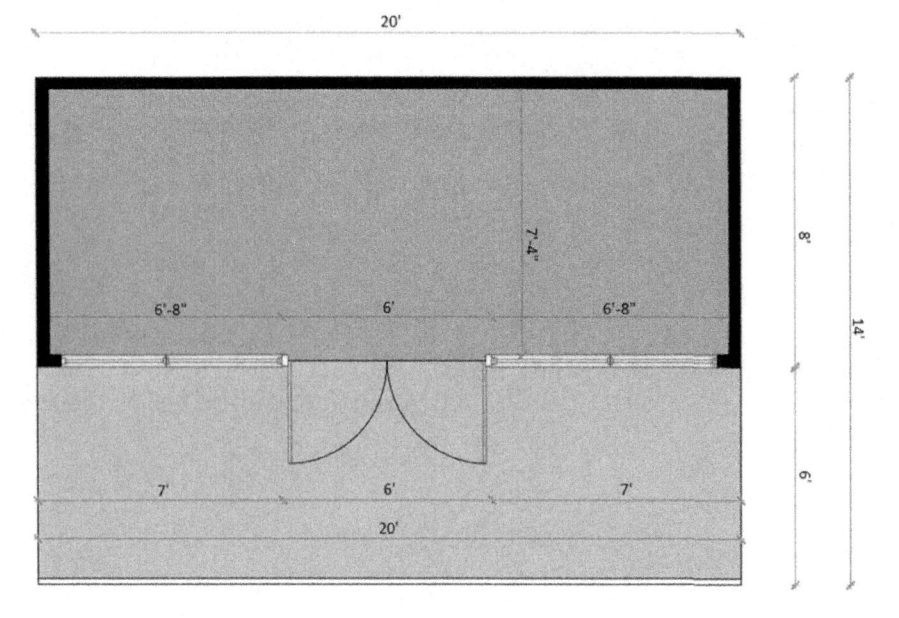

FULL-SCALE FLOOR PLAN IDEAS

Plan 4 has been designed as a hunting lodge from a single 20-foot container. It features a large open-plan living room with a small kitchen and bedroom off to the side. A full-length deck lines the front of the house, making it ideal for relaxing comfortably and taking in the night air. This offers a luxurious space for a single-person dwelling and can offer a comfortable space for 2 if a sofa bed is placed in the living room.

Plan 5

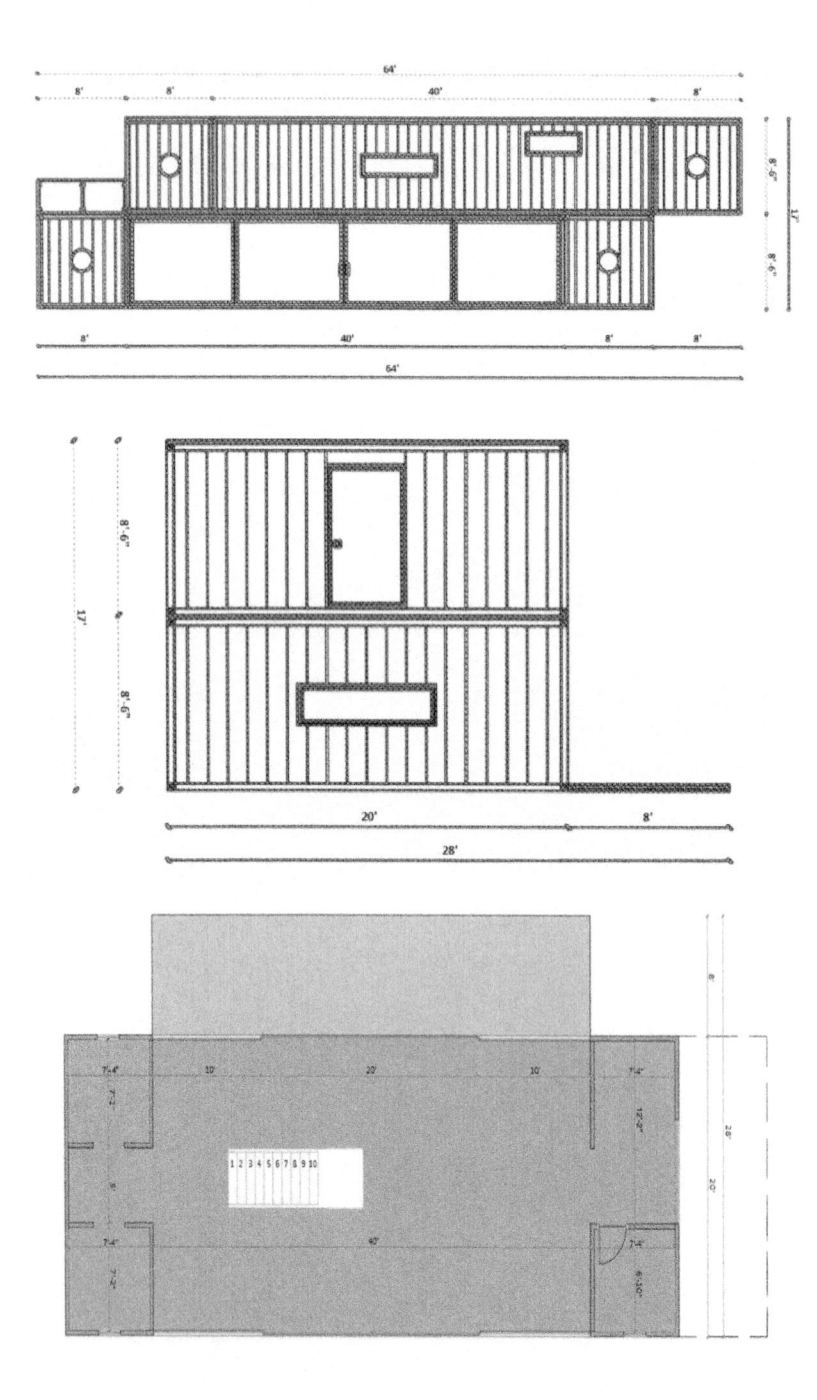

Upstairs:

FULL-SCALE FLOOR PLAN IDEAS

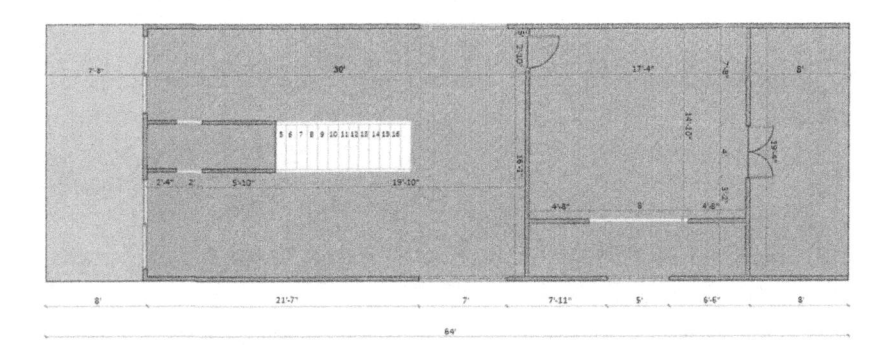

Plan 5 has been designed from 4 40-foot containers and 4 20-foot containers. It demonstrates the luxury possible with shipping container homes. This mansion features 2 floors, 3 bedrooms, 3 bathrooms, and an open-plan living room on both floors. Sliding glass doors on both the front and rear entrance, and a second-floor deck makes this design equal to any traditional home built with the height of decadence.

Plan 6

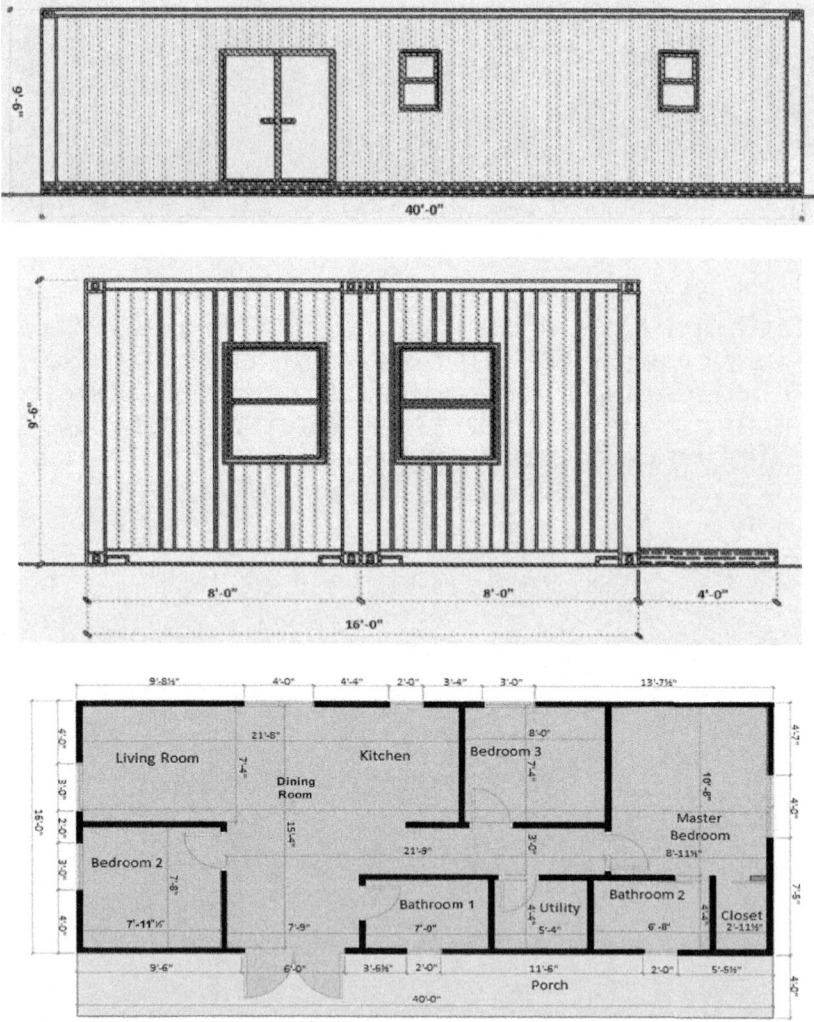

Plan 6 features 3 bedrooms and 2 bathrooms as well as a living room, dining room, kitchen, utility room, and closet. Ample living and storage space, all within 2 40-foot containers. In addition, there is a full-length deck lining the front of the dwelling. With 606 square feet, this dwelling can comfortably house 3 adults or a family of 4.

FULL-SCALE FLOOR PLAN IDEAS

Plan 7

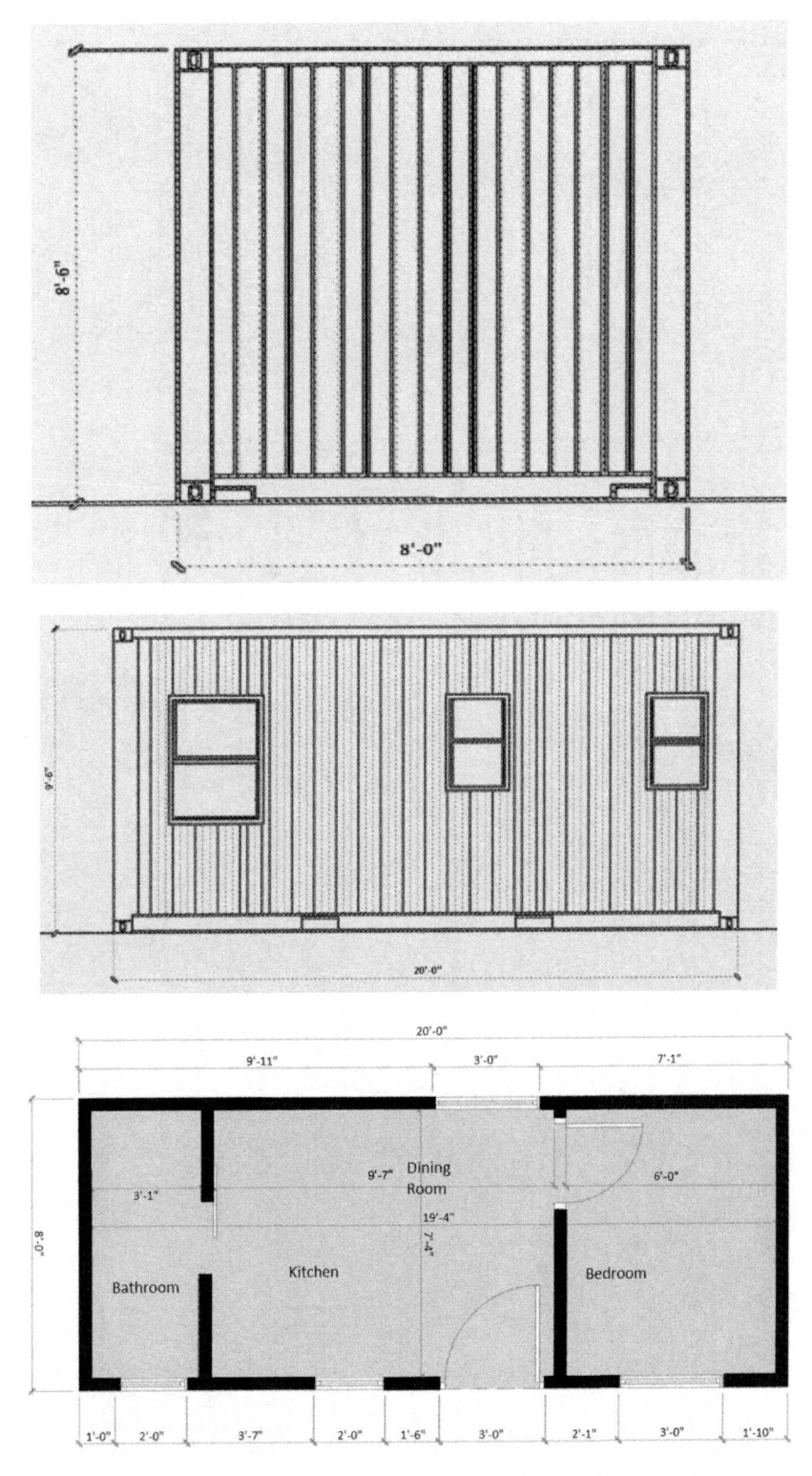

Plan 7 shows another design for a single 20-foot container. This design combines spaciousness with efficiency, featuring a luxurious bathroom and master bedroom with a combined kitchen and dining room. This design is perfect for a single person or a couple, offering all the amenities necessary for a wilderness love nest within 135 square feet.

SHIPPING CONTAINER HOMES

Plan 8

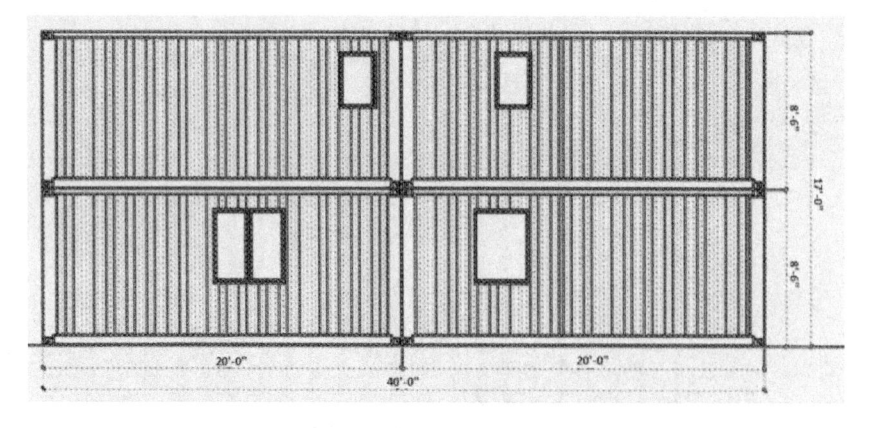

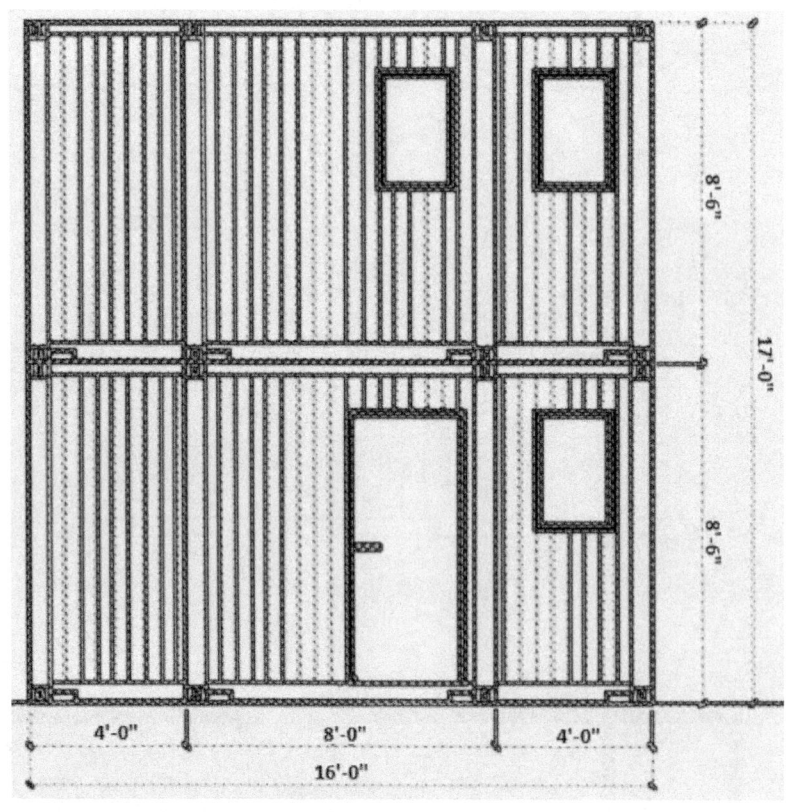

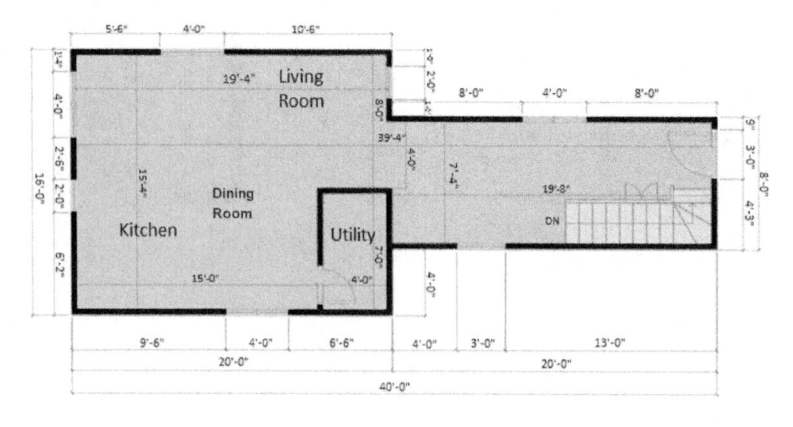

Upstairs:

FULL-SCALE FLOOR PLAN IDEAS

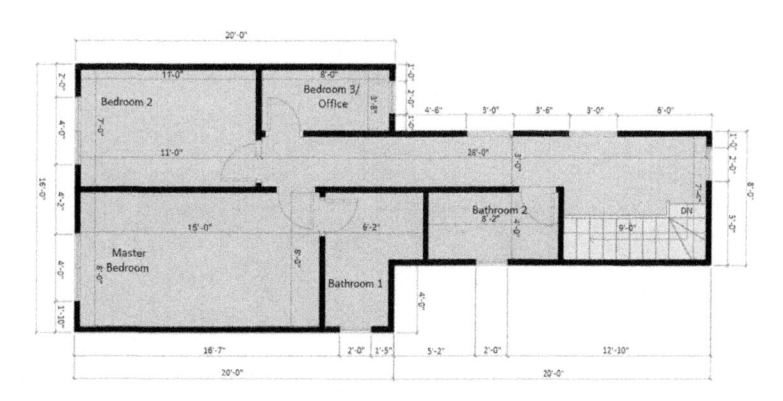

With 6 20-foot containers and 861 square feet of floor space, Plan 8 offers the utmost luxury. It features 3 bedrooms and 2 bathrooms on the upper floor, while the lower floor is devoted to a combined open-plan living room, dining room, and kitchen. Also, included is a utility closet that serves your storage needs.

SHIPPING CONTAINER HOMES

Plan 9

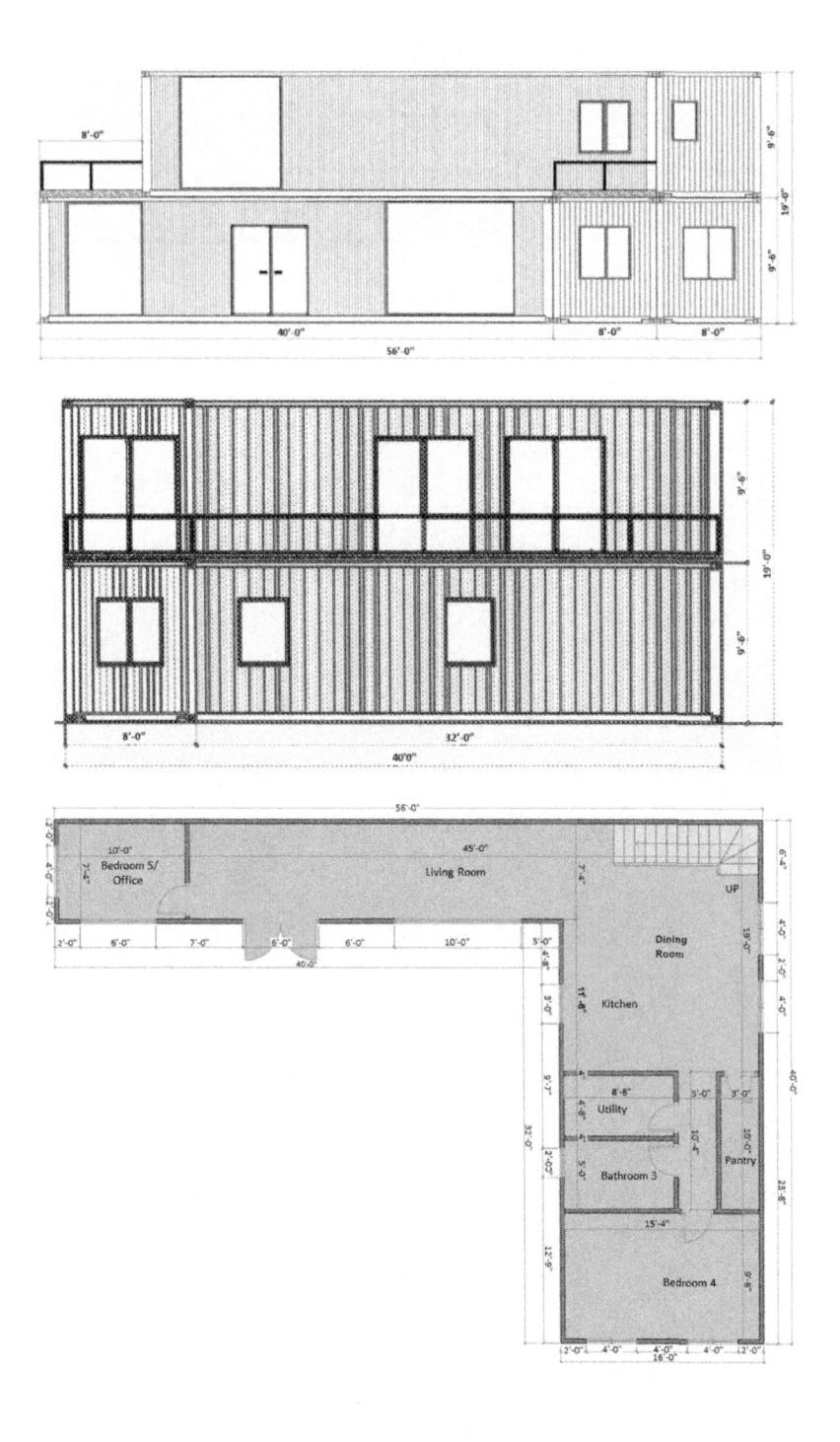

Upstairs:

FULL-SCALE FLOOR PLAN IDEAS

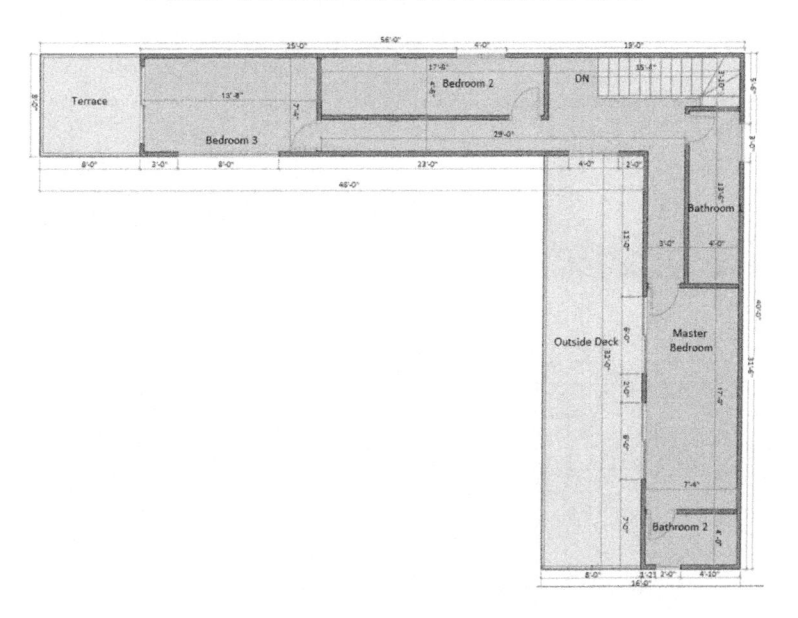

Plan 9 is composed of 5 40-foot high cube containers and provides a total of 1,718 square feet of floor space. It houses 5 bedrooms and 3 bathrooms as well as a combined kitchen and dining room. Also, included are a pantry and utility closet. The second floor features an outside deck and terrace which are perfect for a combination of privacy and fresh air.

SHIPPING CONTAINER HOMES

Plan 10

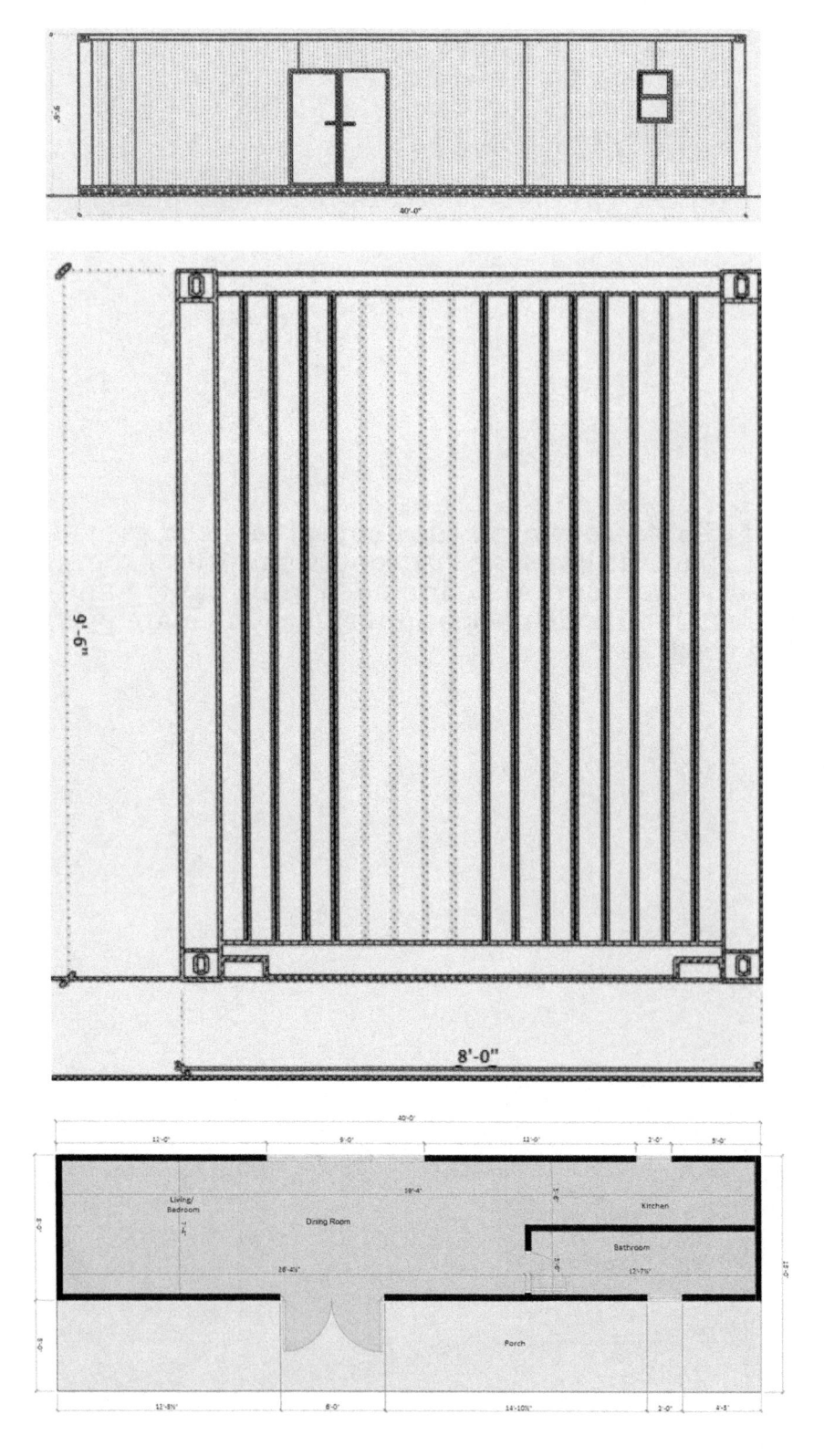

Constructed from a single 40-foot container, Plan 10 offers a stunning 483 square feet of floor space. It features a combined bedroom/living room and open-plan dining room, as well as a spacious bathroom and cozy kitchen. This design is ideal for 2 and can comfortably house 3 or more with a sofa bed.

FULL-SCALE FLOOR PLAN IDEAS

Plan 11

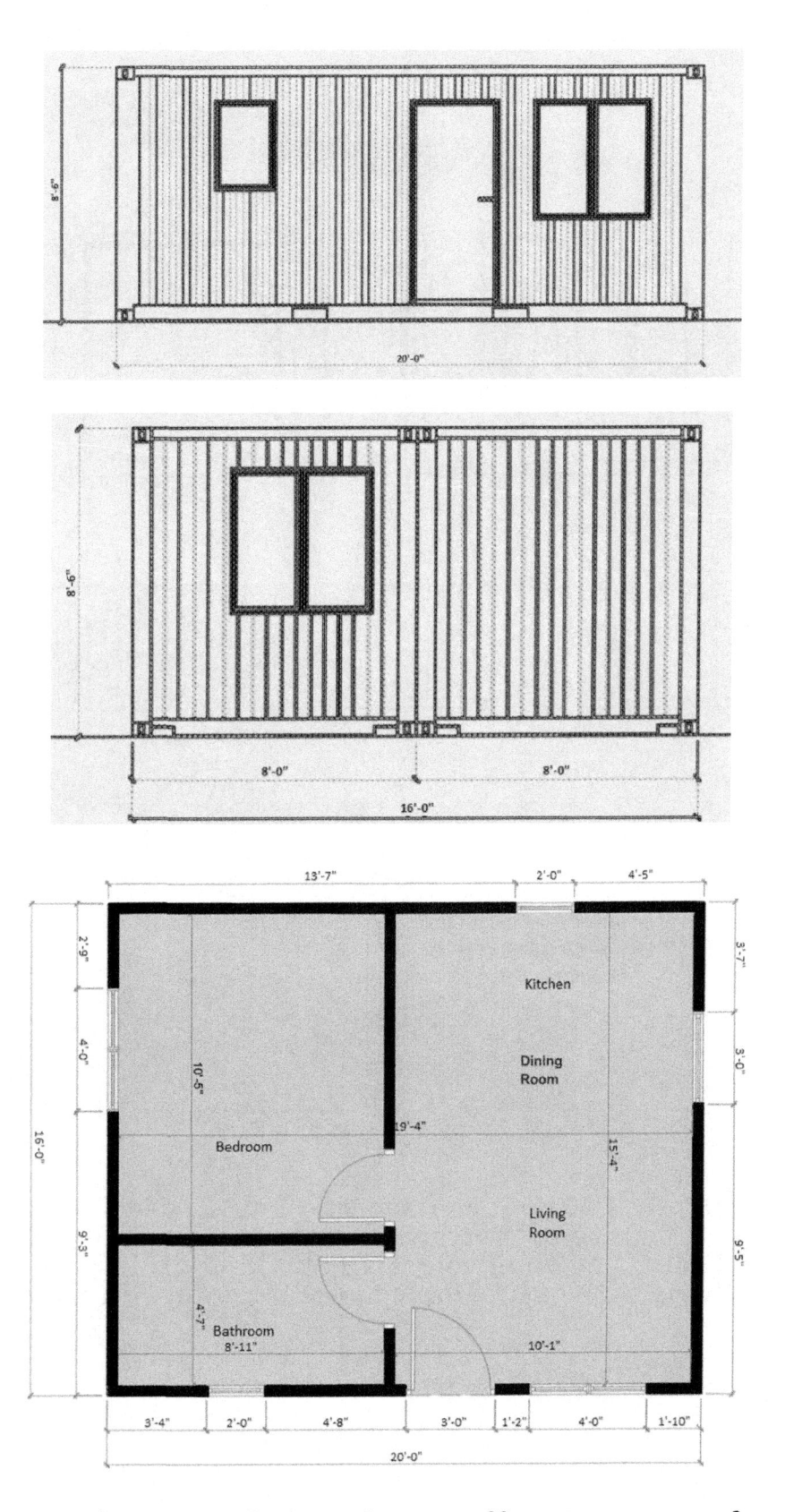

Made from only 2 20-foot. containers, Plan 11 offers 289 square feet of floor space. It features a master bedroom, luxurious bathroom, and combined kitchen and dining room. The front door opens into an open-plan living room which is ideal for visitors, and which can be supplied with a sofa bed to comfortably house a second person.

Plan 12

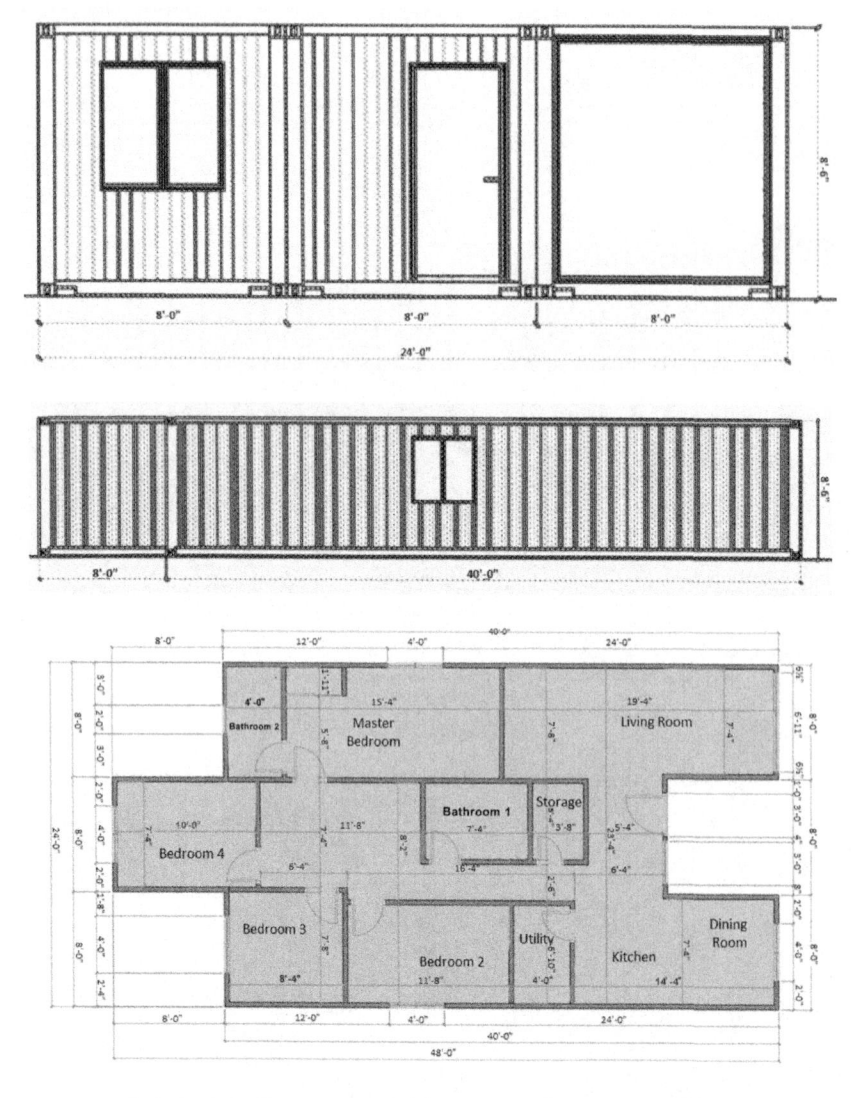

Constructed from merely 3 40-foot containers, Plan 12 offers a spacious 899 square feet of floor space. It features a master bedroom as well as 3 additional bedrooms and 2 bathrooms. It also has an efficiently combined dining room and kitchen as well as a separate living room with ample space for entertaining visitors.

FAQ

FAQ

1. Which States of the US Allow Shipping Container Homes?

Most US states allow them, though the regulations may vary. You should check the rules and regulations of the state before you plan for your shipping container home.

2. Which Countries Allow Shipping Container Homes?

As of today, the US and Canada are the biggest countries in North America that allow for shipping container homes. As for Europe, many European countries have also legalized them like France, the UK, Belgium, Austria, Germany, and Spain. You should, however, always check first or consult with a legal professional before going forward.

3. Are They Safe?

To put it shortly, yes, they are safe. They are made of corrugated steel and they are durable since they are designed to withstand long trips in the sea or on trains, so they are quite durable.

4. Are They Cheaper Than the Average Traditional House?

Yes, they are. However, depending on your budget, you can make a shipping container home almost as expensive as a traditional house. Fortunately, you can make one for a fraction of the cost, too.

5. Can a Shipping Container Home Have a Basement?

Depending on your initial design, you can make room for a basement or a sub-floor to be used as a garage, yes.

6. Can Shipping Container Homes Be Moved?

Yes, they can, and this gives them another edge over a traditional house, which can never be moved.

7. Do You Have to Pay Property Tax for Shipping Container Homes?

If you want your shipping container home to be considered as a piece of real estate (meaning it qualifies for mortgages and can be used as collateral), then you will need to pay property tax.

8. How Much Does an Average Shipping Container Home Cost?

On average, a shipping container can cost between $60,000 and $80,000. However, you can make a more affordable shipping container home for as low as $20,000.

9. Should You Build It Yourself or Hire Contractors?

This will depend on several factors, starting with your budget. You can do it yourself, but you must check a detailed guide, like the one in this book, on how to move forward. If, however, you are not so handy or don't want a DIY approach, then you can use contractors.

10. How Long Can a Shipping Container Home Last?

The average shipping container home can last you up to 25 years without complications. If you regularly maintain it, though, and handle any rust or leakage problems promptly, then it could last much longer.

CONCLUSION

Thank you for making it through to the end of the *Shipping Container Homes*; let's hope it was informative and able to provide you with all of the tools you need to achieve your goals, whatever they may be.

I hope that you found this guidebook to be exactly what you were searching for. As you are now acquainted with all sorts of necessary information related to shipping container homes, you can easily build one. Not only can you use shipping containers for building a home, but you can also use them for any other kind of building purpose. If you are tight on your budget, getting a shipping container home might turn out to be a great idea. Besides everything, this guidebook also provided you with lots of ideas related to flooring, insulation, and also every possible thing that you need to keep in mind while designing your home.

Keep in mind that proper and thorough research can turn out to be your best friend when you are opting for a new type of house. Try to consider the location, weather, and condition of the place that you choose for building your container home. Do not forget to get some proper ideas about the regulations of your area. The key here is to have everything in place so that you can easily avoid any kind of future alterations or issues. Shipping container homes are easily customizable. So, you can unleash your creative side while designing your new home.

I'm positive you now have a clear roadmap to starting your own Shipping Container Homes journey. I wish you all the luck in the world with your **Dream-House**. Do let me know what you think of this book by leaving me a review on Amazon!

Thank You!

Robert Savings

Printed in Great Britain
by Amazon